P9-DUQ-237

INSIDE THE ATOM

Abelard-Schuman books by Isaac Asimov

BUILDING BLOCKS OF THE UNIVERSE (*New Updated 1974 Edition*)
THE CHEMICALS OF LIFE
THE CLOCK WE LIVE ON (*Revised Edition*)
THE DOUBLE PLANET (*Revised Edition*)
ENVIRONMENTS OUT THERE
INSIDE THE ATOM (*New Updated 1974 Edition*)
THE KINGDOM OF THE SUN (*Revised Edition*)
ONLY A TRILLION
RACES AND PEOPLE (*with William C. Boyd*)
THE WORLD OF CARBON

NEW UPDATED 1974 EDITION

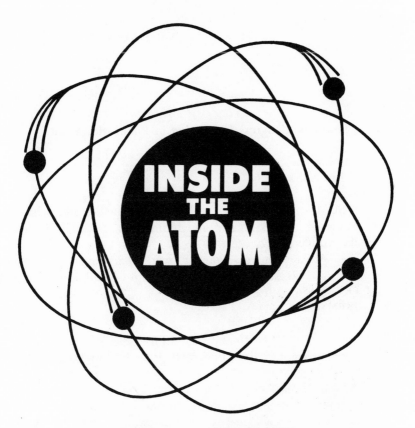

ISAAC ASIMOV

Illustrated by John Bradford

ABELARD-SCHUMAN

LIBRARY

NOV 2 1978

DISCARD

West Hills College

TO CHARLES R. DAWSON

Library of Congress Cataloging in Publication Data

Asimov, Isaac, 1920-
Inside the atom.

SUMMARY: Discusses in detail the structure and behavior of the atom and its present and future uses, especially in light of the world's energy crisis.
1. Atomic energy—Popular works. 2. Atoms. [1. Atomic energy. 2. Atoms] I. Bradford, John Carroll, illus. II. Title.
QC778.A84   1974   539.7   74-5458
ISBN 0-200-71444-9

New Updated 1974 Edition
Copyright © 1956, 1958, 1961, 1966, 1974 by Isaac Asimov

All rights reserved. No part of this book may be reprinted, or reproduced or utilized in any form or by any electronic, mechanical or other means, now known or hereafter invented, including photocopying and recording, or in any information storage and retrieval system, without permission in writing from the Publisher.

Printed in the United States of America.

CONTENTS

ILLUSTRATIONS

INTRODUCTION

In the fall of 1973, the United States was suddenly introduced to the energy crisis, and most people were surprised.

They shouldn't have been. It was inevitable. There is only so much coal and oil in the ground, and it has been the burning of coal and oil that has been mankind's chief source of energy for two centuries now.

In the last generation, mankind has shifted increasingly from coal to oil. Oil is a liquid and it is much simpler to get oil out of the ground than to get coal. Oil is much easier to transport and much easier to use than coal is.

But there is much less oil than coal. At the present rate of burning oil, the oil wells of the world will be pumped dry in thirty years. If we skimp on oil as much as we can, perhaps the oil wells will last fifty years. Even that isn't much time.

We can turn to coal, but coal is difficult and dangerous to mine, hard to transport, not easy to use. We can try to get oil out of a kind of rock called shale—but that is difficult, dangerous and expensive.

Why is all this so desperately important?

Well, the earth now has a population of close to four billion, and the figure is mounting daily. America's population, now approaching 220,000,000, has the highest standard of living of any people in history, and other nations, less fortunate, are trying to raise their own as high. But a record high standard of living for a record number of people is made possible by only one thing: the machine.

Machinery helps grow more food, helps dig into the ground for more ore to convert into more metal, builds and maintains our cities, carries us and our goods over land and sea and through the air, runs our homes, entertains us and keeps us comfortable, even does our mental work for us. If ever our machine civilization were to fail us, even for a short time, starvation and complete disaster would follow. Human muscles alone, without machinery, couldn't support our present society even for a day.

But all our machines depend on energy obtained from the burning of oil or coal, and our oil and coal won't last very much longer. In order to avoid disaster, then, we will need new sources of energy soon to keep our machinery going — energy from uranium fission, energy from hydrogen fusion, energy directly from the Sun. Our machinery is built out of metal and when the supply of certain metals runs low, we must find new sources; we must learn to use low-grade ore or sea-water as a source. Or else we must find substitutes, new ways of using glass, plastics, natural and artificial fibers.

As population continues to increase, we must find new ways of producing and handling food, new ways of fighting insects, weeds, and other pests, new and more efficient ways of housing people and moving them from place to place. Naturally, we all want to be spared the ravages of disease and pain; we want to live longer and be healthier; we want to be protected from flood, fire, and disaster generally.

We must also learn better methods for limiting population; for exploring space; for understanding the workings of the brain. All of this involves science—more science and scientists.

But in order to get more scientists, we must start with young people. It takes time and training to become a skilled research scientist just as it does to become a skilled athlete. In both cases, an early start improves a youngster's chances.

A book such as this one is my way of contributing toward a possible early start. However, that is not the only reason for the book.

It's a mistake to think that there is no point in becoming interested in science if you don't intend to be a scientist someday; or, that there is no longer any point to reading about science if you have finished your schooling.

Consider baseball instead of science for a moment. Baseball is our national sport and there are few of us who don't know the difference between a basehit and a fielder's choice. Yet the percentage of professional baseball players in our population is extremely small.

Would you think that there's no point in being interested in baseball if you're not going to make your living at it? Of course not.

There is, after all, a pleasure in being a spectator, too. If baseball is properly understood, watching two fine teams play a tensely fought game adds to the enjoyment of life. It makes it possible for us to experience thrills and excitement; it gives us hopes and triumphs; yes, and sorrow, too, but a sorrow that is washed out in the thought of a better break tomorrow or even next year.

But without an understanding of the game, we lose all that. If we were to watch a baseball game without knowing the rules or being aware of the fine points, we would see only a group of men chasing a ball.

And so it is with science. Science has become part of our lives and we can't hide from it any longer. It is all about us, touches everything we do. Only a few of us can be research scientists, perhaps, but all the rest of us, whether we like it or not, are spectators.

We can be spectators without understanding and the whole thing will only puzzle and worry us. Or we can learn some of the rules of the game, so to speak, perhaps not enough to make us scientists, but enough to make us appreciative spectators — science-fans, in a way, who know when to be excited and when to cheer.

There is active pleasure in knowing, in understanding. If we do no more with our learning than look at the world about us with more understanding eyes, it will have paid for itself many times over. And there is always the chance that someone will start to learn with only the intention of being an appreciative spectator and end by finding himself part of the game.

Inside the Atom is an attempt to explain some of the fine points of the atom and what goes on inside it; how man has learned about it and what he has done with his learning.

ATOMIC CONTENTS

What All Things Are Made Of

There are so many things in the world that are so completely different from one another that the variety is bewildering. We can't look about us anywhere without realizing that.

For instance, here I sit at a desk, made out of wood. I am using a typewriter made out of steel and other metals. The typewriter ribbon is of silk and is coated with carbon. I am typing on a sheet of paper made of wood pulp and am wearing clothes made of cotton, wool, leather, and other materials. I myself am made up of skin, muscle, blood, bone, and other living tissues, each different from the others.

Through a glass window I can see sidewalks made of crushed stone and roads made of a tarry substance called asphalt. It is raining, so there are puddles of water in sight. The wind is blowing, so I know there is an invisible something called air all about us.

Yet all these substances, different as they seem, have one thing in common. All of them — wood, metal, silk, glass, flesh and blood, all of them — are made up of small, separate particles. The earth itself, the moon, the sun, and all the stars are made up of small particles.

To be sure, you can't see these particles. In fact, if you look at a piece of paper or at some wooden or metallic object, it doesn't seem to be made of particles at all. It seems to be one solid piece.

But suppose you were to look at an empty beach from an airplane. The beach would seem like a solid, yellowish stretch of ground. It would seem to be all one piece. It is only when you get down on your hands and knees on that beach and look closely that you see it is really made up of small, separate grains of sand.

Now the particles that make up everything about us are much smaller than grains of sand. They are so small, in fact, that even a microscope could not make them large enough to see, or anywhere near large enough. The particles are so small that there are more of them in a grain of sand than there are grains of sand on a large beach. There are more of them in a glass of water than there are glasses of water in all the oceans of the world. A hundred million of them laid down side by side would make a line only half an inch long.

These tiny particles that all things are made of are called *atoms*.

How many different kinds of atoms are there? When you think of the millions of different things in the world, you probably suppose there must be millions of different kinds of atoms. That is not so. The number of kinds of atoms known today is exactly 105. That's all. Just 105.

What's more, many of those 105 varieties are very rare. Some occur only in certain uncommon rocks. A few of them are manufactured by scientists and don't exist at all except in laboratories.

In fact, about 99 percent of everything on earth is made up of only about a dozen different kinds of atoms. Such things as sugar, starch, wood, cotton, and vinegar are made up of only three kinds of atoms, the same three kinds in each case. The reason for the variety on earth is that even a few kinds of atoms can be arranged in many different ways. It's as though you were considering threads of only three or four different colors. The number of colors may be small, but the threads can be woven into millions of different designs.

By now you may be asking: If atoms are so small that we can't see them, how do we know that they really exist?

Well, for hundreds of years now, scientists have been trying to determine why some substances burn when they are heated, why some fizz and some explode, why other substances rust in damp weather, and so on. They have experimented in order to find out why the materials that make up the earth behave as they do under different conditions. This type of study is called *chemistry.*

In order to explain the results of their experiments, chemists finally decided that the small particles called atoms exist. Unless there are atoms, there is no simple way of explaining many of the discoveries that chemists have made. This modern *atomic theory* (the ancient Greeks had one, too) was first proposed in 1803 by an English chemist named John Dalton.

In the hundred and fifty years since then, all the experimental evidence has continued to back up this notion. Today it scarcely seems possible to doubt the existence of atoms even though we never see them.

Are atoms the smallest things of all? At first it was thought so. The very word "atom," in fact, comes from a Greek word meaning "uncuttable" or "unsplittable." The idea was that with the atom we had gotten down to rock bottom. It couldn't be cut. It couldn't be split. There was nothing smaller. For nearly a hundred years, chemists thought that.

Then, in the 1890's, scientists studied some of the events that take place when an electric current passes through a vacuum, and they came to the conclusion that particles smaller than atoms do indeed exist. It turned out, in fact, that all atoms are made up of these still smaller particles. These new extra-small objects are called *sub-atomic particles*.

The Two Kinds of Electricity

There is a long history leading up to that discovery in the 1890's. It turns out that people had observed the effects of sub-atomic particles (without knowing it) long before they suspected the existence of such particles. The ancient Greeks, for instance, some 2,500 years ago, noticed that if a piece of amber (a yellowish, glassy substance) is rubbed with fur or cloth, it suddenly becomes able to attract light objects such as small feathers or bits of wool.

Beginning in 1570, an English doctor, William Gilbert, was studying this odd behavior of amber. He found other substances which acted in the same way. Since the Latin word for amber is "electrum," Gilbert called all substances which could be made to show a force of attraction, on being rubbed, "electrics." Soon people came to call that force of attraction *electricity*.

Then, in 1733, a French experimenter, Charles François Du Fay, found there were two kinds of attractive forces; two kinds of electricity.

A glass rod and a rod of sealing wax, he found, if rubbed with silk, both become electrified. The glass rod attracts small objects, and so does the sealing wax.

Suppose, though, that two electrified glass rods are hung by silk threads near each other. When this is done, the two rods swing away from each other. They repel each other. Exactly the same thing happens if two electrified sealing-wax rods are hung near each other. They repel each other, too.

Now suppose that a glass rod and a sealing-wax rod are hung up near each other. These two rods swing closer together. They attract each other.

So, you see, there seems to be one kind of electricity in glass and a second kind in sealing wax. Two objects containing the same kind of electricity, such as two glass rods or two sealing-wax rods, repel each other. Two objects containing different kinds of electricity, such as a glass rod and a sealing-wax rod, attract each other.

Benjamin Franklin, the famous American patriot of Revolutionary days, studied electricity in the 1740's. As a result of his experiments, he suggested that there was one kind of electricity that moved from object to object. Some objects would pile up more than the normal amount of electricity. Others would have less than the normal amount. Franklin was the first to speak of a *positive charge* of electricity and a *negative charge*.

In 1800, an Italian scientist, Alessandro Volta, devised a method of putting metals together in such a way as to construct an *electric battery*. This was a device which could produce a moving current of electricity.

ATTRACTION AND REPULSION

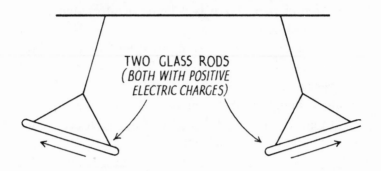

TWO GLASS RODS
*(BOTH WITH POSITIVE
ELECTRIC CHARGES)*

LIKE CHARGES REPEL EACH OTHER

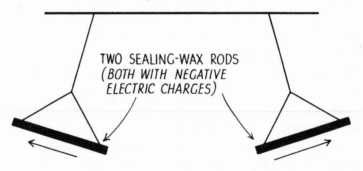

TWO SEALING-WAX RODS
*(BOTH WITH NEGATIVE
ELECTRIC CHARGES)*

LIKE CHARGES REPEL EACH OTHER

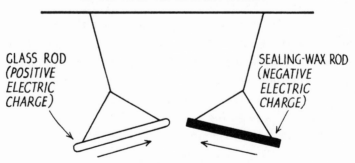

GLASS ROD
*(POSITIVE
ELECTRIC
CHARGE)*

SEALING-WAX ROD
*(NEGATIVE
ELECTRIC
CHARGE)*

UNLIKE CHARGES ATTRACT EACH OTHER

Every battery had a *positive pole,* which was also called an *anode,* and a *negative pole,* which was also called a *cathode.* People working with such batteries guessed that the electric current travelled from the positive pole to the negative pole. They had no way of telling whether they were right; it was just a guess.

The guess turned out to be wrong. As long as the electric current passed through wires or through liquids, it was impossible to tell in which direction it was moving. What if the current were made to pass through a region where there was nothing at all; through a *vacuum,* in other words.

Once scientists had learned how to prepare containers from which almost all the air had been pumped, they were ready to go. Such a *vacuum tube* could be made which contained an anode and a cathode, and electricity could be forced through. It turned out that a glowing stream of electricity originated at the negative electrode, the cathode, and shot straight across the tube in a straight line.

In 1876, a German scientist, Eugen Goldstein, called the rays in this stream of electricity, *cathode rays,* because they started at the cathode. In 1886, he made use of a special cathode in which he had bored holes or "channels." He found that, in using such a cathode, a different set of rays could be made to appear. These passed through the channels and moved in the opposite direction from that in which the cathode rays moved. Goldstein called these new rays *channel rays.*

For years, scientists wondered what these rays might be. Were they a new kind of light, or were they streams of tiny particles? Finally, in 1897, an English scientist, Joseph John Thomson, was able to show that the cathode rays consisted of one kind of very tiny particle; a particle much smaller than any atom. It was the first sub-atomic particle to be discovered. Because an electric current is made up of these particles in motion, they came to be called *electrons.* In 1906, Thomson received a Nobel Prize for this discovery.

The channel rays also consisted of particles; but of particles of different kinds, none of which were electrons. Some of these channel-ray particles were smaller than others. In 1914, a New Zealand-born British scientist, Ernest Rutherford, suggested that the smallest of the channel-ray particles be called *protons.* This suggestion was adopted.

As it turned out, both electrons and protons carried an electric charge. The electron carried a negative charge, and the proton carried a positive charge. They were the particles making up the two kinds of electricity discovered by Du Fay.

However, the electron moved from object to object much more easily than the proton did. When electrons moved from object A to object B, object B was filled with a greater than normal quantity of electrons and carried a negative charge. The more electrons crowded into object B, the greater its negative charge. In the same way, object A, having lost electrons, was left in the opposite condition and carried a positive charge. The more electrons left object A, the greater its positive charge.

If an object carrying a negative charge touched one carrying a positive charge, the extra electrons in the first object flowed into the second object to make up its deficit. The electrons level out in this way, and the objects are *discharged.* Sometimes the electrons force their way from one object to the other through the air just before they touch. Then there is a little spark of light and a sharp crackle.

During thunderstorms the ground and the clouds carry electric charges; very large ones. When an electric dis-

CATHODE RAYS
AND
CHANNEL RAYS

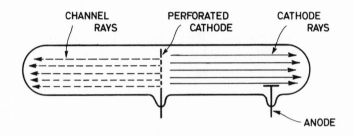

CHANNEL
RAYS

PERFORATED
CATHODE

CATHODE
RAYS

ANODE

CHARGE
AND
DISCHARGE

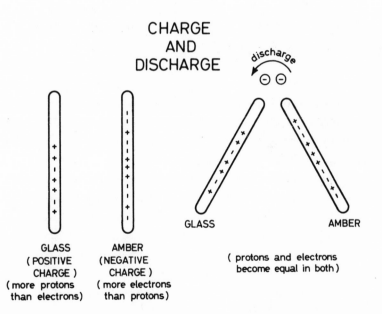

discharge

GLASS

AMBER

GLASS
(POSITIVE
CHARGE)
(more protons
than electrons)

AMBER
(NEGATIVE
CHARGE)
(more electrons
than protons)

(protons and electrons
become equal in both)

charge takes place during such a storm, the spark of light is a bolt of lightning, and the sharp crackle is a crack of thunder. It was Franklin who first showed that in 1752, when he flew a kite in a thunderstorm and brought some of the electricity down to earth.

But why should an object gain a positive charge if electrons leave it? The answer to that is that all atoms contain both electrons and protons; particles of negative charge and particles of positive charge. Ordinarily, matter contains equal numbers of both, so that the effect of one charge just cancels the effect of the other. Ordinary matter is *uncharged*.

If electrons enter an object, that object contains more electrons than protons and the negative charge overbalances the positive and can be detected. If electrons leave an object, that object contains more protons than electrons, and now it is the positive charge that overbalances and shows up.

The Unequal Twins

Both electrons and protons are much smaller, very much smaller, than atoms. It takes one hundred thousand electrons or protons lying side by side to stretch across the space taken up by a single atom.

All atoms contain within themselves protons and electrons, at least one of each. Some atoms have as many as a hundred and three of each. Both protons and electrons are therefore examples of the sub-atomic particles we mentioned earlier.

Of course, electrons and protons are different in the kind of electric charge they carry. We have already mentioned that. There is another important difference, too, a difference that showed up when electrons and protons were studied in connection with magnets.

You are probably quite familiar with small magnets that

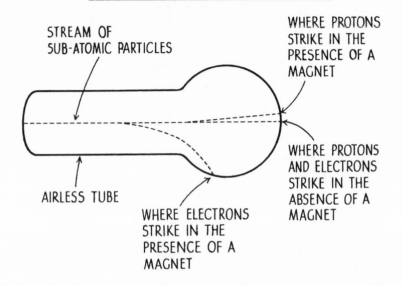

SUB-ATOMIC PARTICLES AND MAGNETS

STREAM OF
SUB-ATOMIC PARTICLES

WHERE PROTONS
STRIKE IN THE
PRESENCE OF A
MAGNET

AIRLESS TUBE

WHERE PROTONS
AND ELECTRONS
STRIKE IN THE
ABSENCE OF A
MAGNET

WHERE ELECTRONS
STRIKE IN THE
PRESENCE OF A
MAGNET

can lift up needles, pins, and other little objects made of iron and steel. The simplest magnet is just a bar of magnetized steel. If such a magnet is free to swing horizontally in any direction, as a compass needle is, it will point north and south, as a compass needle does. The end that points north is called the magnet's *north pole*, and the other is the *south pole*. The common horseshoe magnet is simply a bar bent so that the north pole and the south pole are next to each other.

If you have two straight-bar magnets, the north pole of one and the south pole of the other will attract each other. The magnets will come together with a clank and remain stuck together. The north pole of one, if shoved toward the north pole of the other, will resist the movement. You will have to use force to make them come together, and even then they won't stick. The same is true of two south poles. Here again, as in the case of electricity, two unlikes will attract each other while two likes will repel each other. (Electricity and magnetism are very closely related, and you can't have one without the other.)

Moving electrons and protons are affected by magnets. The electrons are pulled in one direction, the protons in the opposite direction. That, indeed, is how you can tell the two kinds of particles carry opposite charges. From the particular direction in which they are pulled by the magnet you can tell that it is the electron that carries the negative charge and the proton the positive one. Thus, if a stream of electrons is passed through an airless glass tube, a bright spot appears at the end of the tube, where the electrons strike the glass. If a magnet is brought near the stream, the electrons move away from their usual straight line of travel, and the bright spot moves, too.

The amount by which the electron stream is shifted from its straight path depends partly on how heavy the electrons are. Just imagine kicking a billiard ball as it rolled by; you would change its direction completely. Suppose, though, it was a cannon ball (moving no faster than the billiard ball) that you kicked. While you were hopping on one foot (because you had hurt the one you kicked the ball with), you would notice that the heavy cannon ball had changed direction very little. In the same way scientists noticed that moving protons change direction much less under the influence of a magnet than moving electrons. The proton, they decided, is obviously a much heavier particle than the electron. By comparing the proton shift with the electron shift, they decided that a proton is as heavy as 1,836 electrons.

Now let's stop for a moment. We talk about "heavy" things, about objects that "weigh" something. Weight is the result of the attraction of the earth's gravity. You weigh 120 pounds, let us say, because the earth pulls you with that much force. The moon, a smaller body than the earth, has less gravity, only a sixth as much as the earth has. If you were on the moon, it would pull you less strongly; you would weigh only 20 pounds. On Jupiter, which is much larger than the earth, you would weigh 300 pounds.

You don't even have to go to another planet to change your weight. If you've ever gone swimming, you know that your body feels much lighter in water. Water has a buoyant effect; it lifts you up. If you stretch out on the water, you can usually float without any trouble. When you do that, you actually have no weight at all.

With weight such a changeable thing, confusion might set in. Scientists therefore talk about the weight of an object under certain definite conditions. They consider its weight in a vacuum at sea level at a latitude of 45 degrees.

The weight of any object under these definite conditions is equal to its *mass*. A piece of wood with a mass of one pound weighs exactly one pound when weighed in a vacuum at sea level at 45 degrees north or south latitude. If it were surrounded by air or were on a high mountain, it would weigh a trifle less than one pound. If it were at the North Pole, it would weigh a trifle more than one pound, and at the Equator, it would weigh a trifle less. It would weigh nothing at all if it were floating in water. It would weigh two and a half pounds on Jupiter, twenty-six pounds on the sun, less than three ounces on the moon. Its mass, however, would always be one pound because that's what it would weigh under the definite conditions scientists have decided on.

Scientists use very delicate instruments to weigh small objects, but they don't rush down to a particular spot on Earth to weigh them. They don't insist on being at sea level, and they don't usually try to weigh the objects in a vacuum. They know how to calculate the small differences that result from not having the conditions they have decided on, and when it is necessary to be very accurate, they make allowances for those differences.

Throughout this book, we will always speak of the mass of an object, rather than of its weight. Instead of saying, for instance, that a proton weighs as much (or is as heavy) as 1,836 electrons, we will say that a proton has as much mass (or is as massive) as 1,836 electrons.

Well, then, how much mass does a proton have in pounds? As you can imagine, very little indeed. It takes about 270,000,000,000,000,000,000,000,000 protons to make up a pound of mass. This is a very large number and an inconvenient one. In words, it is two hundred and seventy million million million million, or two hundred and seventy trillion trillion, or simply two hundred and seventy septillion. It would be ridiculous to try to use numbers so large.

Instead, let us agree to call the mass of a proton simply one — the numeral 1. The masses of other sub-atomic particles or even of whole atoms can then be expressed according to the way they compare with the mass of a proton. For instance, an atom that had as much mass as ten protons would have the *mass number* of 10; if it were as massive as seventy-two protons, it would have the mass number of 72; and so on. Atoms can have mass numbers from 1 to over 250.

An electron, as you know, must have a mass number much less than 1, since its mass is much less than that of a proton. In fact, the mass number of an electron is so small (1/1,836) that it is often disregarded altogether.

Since the proton is much more massive than the electron, it would be natural to think that it carries more electric charge. That, however, is not so. The amount of electric charge contained in the proton is exactly the same as the amount of electric charge contained in the electron. The proton contains a positive charge and the electron contains a negative charge, but that is the only difference.

What is the amount of electric charge on a single electron or proton? How much actual electricity do such small particles carry?

If we tried to express this amount in ordinary quantities, we would have to use very inconvenient numbers. It takes all the electricity of millions of trillions of electrons to light up a small bulb for even a fraction of a second. Scientists simplify matters by agreeing to call the electric charge of a proton or electron simply one — the numeral 1. The charge in a proton is positive, so its electric charge is said to be $+1$. The electric charge of an electron, naturally, is -1.

It is because the electron has so little mass that it can move from place to place easily. The proton, being more massive, is also more sluggish. It tends to stay put. The electrical effects we are familiar with, from doorbells to television, are all due to moving electrons.

Some substances allow electrons to pass through very easily. Such substances are called *conductors*. One of the best conductors is copper, and that is why most electrical wiring is of that metal. Other substances do not allow electrons to pass through, or do so only with great difficulty. These are called *insulators*. Common examples of insulators are rubber, silk, wax, glass, and sulfur. Copper wires are often covered with rubber or silk so that they may be handled safely while electric currents are passing through them. The electrons cannot move through the insulators to enter your body and injure it.

The Uncharged Particle

Protons and electrons, which were discovered in the 1890's, are only two of the three important types of particles that make up atoms. It wasn't until 1930 that the third type of particle was discovered.

The reason for the delay was that the third particle lacks a charge. It is the electric charge on the electrons and protons, the way they act under the influence of magnets, that makes them easy to study.

Let's consider a couple of ways in which scientists study the motion of particles, for instance. Imagine a container full of humid air, with a piston on top that can be pulled upward. When the piston is pulled upward, the air expands; and when air expands, its temperature falls.

As the air cools down, it can hold less water vapor than before and some of the vapor settles out as tiny droplets. Those droplets, however, must form about something, usually a tiny fragment of dust.

If no dust is present in the air, the droplets may form about atoms or molecules that carry an electric charge. Such charged atoms or molecules are called *ions*. Ordinary air doesn't have very many ions in it, but when a proton or electron goes charging through, it leaves a trail of such ions in its path.

If the air in the chamber is dust-free and if it is made to expand just as the proton or electron passes through, then water droplets will form on the ions. Those droplets will

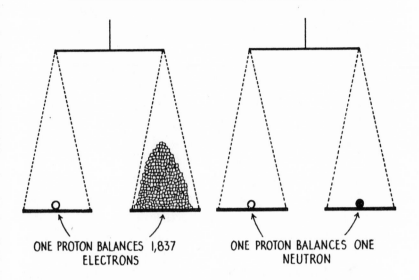

MASSES OF SUB-ATOMIC PARTICLES

ONE PROTON BALANCES 1,837 ELECTRONS

ONE PROTON BALANCES ONE NEUTRON

mark out the path of the speeding particle, which can then be photographed and studied.

Such an instrument, in which humid air forms a tiny cloud that settles out as a line of water drops, is called a *cloud chamber*. It was invented by a Scottish physicist, Charles T. R. Wilson, in 1911. Wilson won the Nobel Prize in physics for that in 1927.

A similar instrument was invented in 1952 by an American physicist, Donald A. Glaser. He made use of a liquid that was hot enough to boil but that was kept under pressure in a closed chamber to keep it from boiling. If the pressure was released, it would start boiling at once, bubbles of vapor forming. These bubbles formed about ions very easily.

If a speeding sub-atomic particle passed through the liquid just before the pressure was released, a trail of ions would be formed. The bubbles would form about the ions, and there would be the track of the particle clearly visible. This *bubble chamber* earned a Nobel Prize for Glaser in 1960.

If the chambers are placed near magnets, the charged particles will move in a curve, so that the line of drops or bubbles will curve. A negatively-charged particle like the electron will curve in one direction, and a positively-charged particle like the proton will curve in the other direction. A light particle will leave a thin, sharply-curved trail. A heavy particle will leave a thick, gently-curved trail.

A lot of information about the particles can be obtained from such trails. An uncharged particle forms no ions when it speeds through matter. For that reason, its passage cannot be detected by a line of water droplets or a line of bubbles. In fact, all instruments that detect sub-atomic particles do so by the ions they form. None of them work for an uncharged particle. That is what kept it hidden so long.

You may wonder, then, how it came to be discovered at all. The answer is that scientists made some observations that couldn't be explained, unless it was decided that an uncharged particle existed.

To see what I mean, imagine you are watching a juggler manipulate his Indian clubs on a darkened stage. If some of his clubs were painted with red luminous paint and some with green luminous paint, they would resemble the electrons and protons in a way. They would be noticeable because of their paint. If the juggler were also using a few unpainted clubs, these would be invisible in the darkness.

But suppose one of the unpainted Indian clubs finally slipped from the juggler's grasp and hit you on the head. Then, at last, you would suspect something was there even though you couldn't see it.

In the early 1930's, it was found that matter could be

made to produce a kind of radiation that could not be detected. If a cloud chamber was placed in the radiation, nothing showed up.

Yet there had to be something there, for if paraffin was placed in the path of the radiation, protons were sent streaking out of the paraffin. Those protons could be detected without trouble. Something had to be knocking those protons out of the paraffin. Since the protons were pretty massive and wouldn't budge for anything as light as an electron or a light wave, that something had to be pretty massive, too.

Finally, an English physicist, James Chadwick, in 1932 announced that the way to explain this was to suppose that the mysterious radiation contained uncharged particles about as massive as protons. These uncharged particles he called *neutrons.* For this discovery he received a Nobel Prize in 1935.

The word "neutron" arises from the fact that the new particles were "neutral." They have neither a positive charge nor a negative charge; they are neither one nor the other. The neutron has almost exactly the same mass as a proton; so its mass number is 1.

We can make a little table, now, describing the different properties of the three types of particles that are found within atoms:

	Mass Number	*Charge*
Proton	1	+1
Electron	0 (almost)	−1
Neutron	1	0

To summarize, then, the electron, the proton, and the neutron are among the smallest things known to exist. Everything in the universe is made up chiefly of these three types of particles. In the next chapter we will see how these particles are arranged inside the atom and how the different arrangements result in different types of atoms.

ATOMIC ARRANGEMENTS

The Massive Center

In 1906 a British scientist named Ernest Rutherford was studying the effect a stream of massive sub-atomic particles had on a photographic plate when they struck it. When the plate was developed, there was a dark spot where the particles had struck. He then placed a sheet of gold foil only one fifty-thousandth of an inch thick in the path of the particle stream. The dark spot was practically unchanged, as if all or almost all the particles had passed right through the gold foil without being bothered by it. But there was some very faint darkening of the negative all around the spot, as if a few, a very few, of the particles had been turned aside by the gold so that they struck the plate in a new place.

There seemed to be only one way to explain this. Thin as the gold leaf was it still contained a thickness of two thousand atoms through which the alpha particles had to pass. Rutherford decided that the gold atoms must be mostly made up of very light particles that could not stop a massive particle from passing through. He decided that the main mass of the atom must all be concentrated in a small spot in the center.

To see what this means, imagine a number of small lumps of lead hanging in air and well separated from one another. Imagine throwing metal pellets at the lumps of lead without aiming. Most of the pellets would simply pass through the air between the lumps without being affected in any way. Occasionally, though, a pellet might hit one of the lumps of lead. It would then bounce off and change its direction of travel.

This is like the situation in the atom. The massive particles are all packed together tightly in the very center. The resulting group of particles is called the *atomic nucleus* (plural, *nuclei*). All the rest of the atom is occupied by the very light electrons. The nuclei of neighboring atoms in any solid substance are separated from one another by electrons, just as our lumps of lead were separated by air. Speeding sub-atomic particles usually pass through the electron regions of the atom without being bothered. Only one out of many thousands will happen to hit the small nucleus in the center, and that one will then bounce off and move in a new direction.

You will remember that sub-atomic particles are very small compared with the whole atom. This means that the atomic nucleus takes up a very small fraction of all the room inside the atom. The most complicated atoms have about 250 particles in the nucleus altogether. Even that many, when packed together, form an atomic nucleus which is so small that it would take 7,000 of them, side by side, to stretch across the space taken up by a single atom.

This means that, if an atom were as large as a basketball, its nucleus would still be only 1/500 of an inch through. It would still be too small to see with the naked eye.

In spite of the small amount of room taken up by the atomic nucleus, almost all the mass of the atom is contained there. Fully 99.95 percent of the mass — and sometimes even more — is found right there in the atomic nucleus.

For working out the notion of the atomic nucleus and for

SUB-ATOMIC PARTICLES
PASSING THROUGH ATOMS

PLANETARY ELECTRONS

ATOMIC NUCLEUS

ATOM

PHOTOGRAPHIC
PLATE

MOST PARTICLES
PASS THROUGH ELECTRONS
AND JUST KEEP ON GOING

STREAM OF
SUB-ATOMIC
PARTICLES

AN OCCASIONAL PARTICLE
HAPPENS TO HIT AN
ATOMIC NUCLEUS AND
BOUNCES OFF TO
ONE SIDE

"SOLID" PIECE OF
METAL FOIL

other important work on atoms, Rutherford received a Nobel Prize in 1908.

At first, very little was known about how the nucleus of one kind of atom differed from that of another. A major discovery was made in 1913 by an English physicist, Henry G. J. Moseley. He was able to show that each nucleus carried a positive electric charge. This was the first indication that the nucleus must contain protons. (Moseley would undoubtedly have won a Nobel Prize for this, but he joined the British Army in World War I and was killed in action in 1915.)

The number of protons that had to be present in the nucleus to account for its electric charge was not great enough to account for its mass. Once the neutron was discovered in 1932, a German physicist, Werner Karl Heisenberg, at once suggested that they, too, were present in the nucleus.

Once it was understood that the nucleus was made up of protons and neutrons, it was easy to work out the mass and electric charge of the nucleus in terms of these particles.

Obviously, the mass of an atomic nucleus depends on the number of protons and neutrons that it contains. Since the mass number of each proton and each neutron is 1, it is only necessary to add up their total number to get the mass number of that particular atomic nucleus. An atomic nucleus which contains two protons and two neutrons has the mass number 4. One which contains eight protons and eight neutrons has the mass number 16. One which contains 92 protons and 146 neutrons has the mass number 238.

The electric charge of the atomic nucleus is just as simple to figure or perhaps even simpler. The neutrons have no charge at all, so they can be ignored. Each proton in the nucleus, however, has a charge of $+1$. The total charge of the nucleus is therefore equal to the number of protons it contains. The nucleus with two protons and two neutrons has a total charge of +2. The one with eight protons and eight neutrons has a charge of +8, and the one with 92 protons and 146 neutrons has a charge of +92.

The number of protons in the atomic nucleus is called the *atomic number*. Be particularly careful to notice that the atomic number and the mass number are two different things. The mass number describes the mass of an atomic nucleus and is equal to all the particles, both protons and neutrons, that the nucleus contains. The atomic number describes the electric charge of an atomic nucleus and is equal to the number of protons only that the nucleus contains.

If you know both the atomic number and the mass number of a nucleus, you can figure out exactly the number of protons and neutrons it must contain. Suppose you were told that a certain nucleus has atomic number 20 and mass number 42. Well, then, since its atomic number is 20, the nucleus must contain twenty protons. To have mass number 42, the nucleus must contain twenty-two neutrons in addition to the twenty protons. And that is all there is to that.

The Frothy Remainder

All the rest of the atom, outside the tiny central nucleus, is made up of electrons. These electrons within the atom are sometimes called *planetary electrons*. The reason for that name is that they were first pictured as little particles whirling about the central nucleus much as the planets revolve about the sun. Actually, things aren't quite that simple. Modern theories of electron motion involve some complicated mathematics, which we don't have to worry about.

For our purposes, the somewhat old-fashioned notion of electrons whirling about the nucleus like planets about a sun is close enough.

The number of electrons in an ordinary atom is equal to the number of protons in the nucleus of that atom. It is equal to the atomic number, in other words.

To go back to our usual examples: An atomic nucleus with two protons and two neutrons would be surrounded by two electrons. An atomic nucleus with eight protons and eight neutrons would be surrounded by eight electrons, and one with 92 protons and 146 neutrons would be surrounded by 92 electrons.

Now let us consider the charge of these electrons. The total charge of the electrons within an atom depends upon their number. Each electron has a charge of −1. Two electrons would have a total charge of −2, eight electrons a charge of −8, and 92 electrons a charge of −92.

The number of electrons in an ordinary atom is equal to the number of protons in the nucleus, as we have seen. That means that the negative charge in the outer regions of the atom is exactly equal to the positive charge on its nucleus. Therefore the total charge of such an atom is exactly zero. That atom contains plenty of both positive and negative electricity, but there are equal quantities of each so that one neutralizes the effect of the other. The atom as a whole is uncharged. It is a *neutral atom*.

What about the mass of the electrons? It is very little indeed. The most complicated atom we know contains 102 electrons. The total mass of all those electrons together is only a little over a twentieth of the mass of a single proton or neutron. That is why people sometimes say that "an atom is mostly empty space."

Don't think for one moment, though, that this means the electrons are unimportant. Electrons are very far from really being empty space. In the first place, their charge neutralizes the charge on the nucleus. Secondly, although they cannot stop speeding sub-atomic particles, they can protect the nucleus from other atoms. When two atoms collide at ordinary speeds, they bounce away again after getting no nearer than each other's outermost electrons.

In the process, though, the outermost electrons in an atom must take some punishment. As a result of heat, certain kinds of atomic collisions, or other causes, one, two, or even three of the outermost electrons can be chipped off an atom.

Under very special circumstances, the inner electrons also can be removed. This happens, for instance, in the interior of stars such as our own sun, where the temperature rises to millions of degrees. There, all the electrons are stripped off the atom. (It wasn't until quite recently that mankind could duplicate that effect.)

In those stars where atoms are stripped of all their planetary electrons, the bare atomic nuclei can approach one another far more closely than the original atoms could. The planetary electrons are no longer there to act as protection or as "bumpers." Such bare nuclei are called *collapsed matter* because matter composed of them can collapse together until the tiny nuclei are nearly touching.

Collapsed matter takes up very little space. Suppose all the atoms in the whole earth were stripped of their electrons and the bare nuclei were allowed to collapse together till they were touching. The earth would be squeezed into a ball a little over a mile in diameter.

Yet the nuclei contain just about all the mass of the atom.

That means that collapsed matter would have all the mass of the original matter even though it took up much less room. The little ball of matter, one mile across, into which we imagined all of the earth to have been collapsed, would be just as massive as all of the earth originally was. A piece of collapsed matter as big as a grain of sand would contain hundreds of tons of mass.

The sun and most ordinary stars (each of which is millions of times as massive as our small planet) contain only a small quantity of collapsed matter in their interiors. Some unusual stars, however, called *white dwarfs,* are made up of collapsed matter almost entirely. They are called "dwarfs" because they are unusually small for stars. Some of them are smaller than the earth. Yet for all their small size they are as massive as other stars.

Some stars collapse all the way, until all the matter in them is turned into neutrons that are squashed into contact. These are *neutron stars,* which are only ten miles across and yet are still just as massive as other stars.

You see, then, that the reason the matter about us is light and fluffy is entirely because of the electrons in the atom. They may be very light but they take up lots of room and keep the massive atomic nuclei well apart.

Varieties of Atoms

Atoms are usually associated with one another in groups called *molecules.* Some molecules are quite small. The air, for instance, is made up mostly of molecules containing two atoms apiece. Larger molecules also exist. Some of the molecules in our body are made up of thousands of atoms.

Molecules are always in motion. The molecules in the air about us, for instance, are moving at speeds close to sixty miles an hour. Even in solid matter, which seems to be hard and motionless, molecules are vibrating rapidly back and forth. The higher the temperature, the faster they move.

Naturally, molecules moving like this collide with one another frequently. There may be millions and billions of collisions every minute. Colliding molecules often bounce apart without being in any way affected. Sometimes, however, changes result. A colliding molecule may have an atom or two knocked out of it, or it may exchange atoms with the molecule it strikes. It may stick to the molecule with which it collides, forming a new and larger molecule. Any of a number of other things may happen.

Such events are often quite visible to us. We may not see the colliding molecules, but we can see the things that happen as a result. The molecules in an antacid tablet collide with one another and with the molecules in water; as a result there is a fizzing, and bubbles are formed. The molecules in a piece of iron collide with the molecules in the air, and as a result we see the iron turn rusty. The molecules in an acid collide with molecules in a piece of copper, and the copper turns green.

When the temperature is raised, and molecules move more quickly, the collisions between them take place more frequently and with greater force. Visible changes take place faster. Paper bursts into flame. Wood chars and burns. Dynamite explodes.

All these changes that result from colliding molecules are called *chemical reactions*.

Chemists study these chemical reactions and try to figure out what different molecules will do in different types of collisions. From this they learn about the nature of the atoms that make up the molecules. It turns out that the way an atom behaves inside a molecule depends on how many electrons it has. If two atoms have the same number of electrons, they will behave in the same way.

As we know, the number of electrons in the neutral atom is equal to the number of protons in the nucleus of that atom. It is therefore equal to the atomic number. For that reason, chemists divide up atoms according to their atomic number. All atoms with the same atomic number behave alike. Atoms with different atomic numbers behave differently.

The simplest atom has atomic number 1. The most complicated atom we know today has atomic number 103. Atoms with all the numbers in between are known. This means that for the chemist there are exactly 103 different kinds of atoms. Chemists have given names to these 103 kinds of atoms, and they are referred to as the *elements.*

Sample Elements

Suppose we take the simplest atom. It has atomic number 1; that is, it has a single electron and no more. The name chemists give to such atoms is *hydrogen.* (The name "hydrogen" comes from two Greek words meaning "to give rise to water." The reason for this name is that, when hydrogen burns, water is actually formed. Most elements, and most other chemicals as well, usually have names derived from Greek or Latin. The names often describe something about the behavior of the chemical or where it was first discovered or some other fact about it.)

Hydrogen atoms pair off to form *hydrogen molecules.* If a large number of hydrogen molecules are collected in one place, the result is an airlike substance called *gas.* Hydrogen is the lightest gas known. It is only a fifteenth as heavy as air; so, when hydrogen is collected inside a balloon, the balloon floats in air just as wood floats in water. Balloons or dirigibles can be made large enough to carry tons of mass into the air.

Hydrogen has one troublesome characteristic, though. It has a tendency to get itself involved in chemical reactions. In particular, the collision of hydrogen molecules with certain molecules in the air can result in considerable activity. If the temperature is high enough, hydrogen will react so quickly that it will explode. The hydrogen in the giant dirigible *Hindenburg* exploded back in 1937. That spectacular disaster was probably caused by the heat produced by a spark of static electricity. Because hydrogen enters into chemical reactions in this way, it is called an *active element*.

The next-simplest atom, of course, is the one with atomic number 2. It contains two electrons. This atom is called *helium*. (The word "helium" comes from the Greek word for "sun" because, believe it or not, that element was discovered in the sun before it was discovered here on the earth.)

Helium is also a gas that is lighter than air. It is not as light as hydrogen, but it is still light enough to use in dirigibles. The useful thing about helium is that it gets involved in no chemical reactions at all. Helium atoms can collide with any molecule and just bounce off unchanged, even at high temperatures. They don't even group themselves with one another. They remain single atoms. Helium won't burn; it won't explode; it won't anything. It is an *inert element*. It is therefore quite safe to use in dirigibles. If you have ever seen the kind of small floating balloons sold at fairs and parades, you can be sure they have been filled with helium.

You may be surprised that a single electron can make so much difference, but it does. Hydrogen, with one planetary

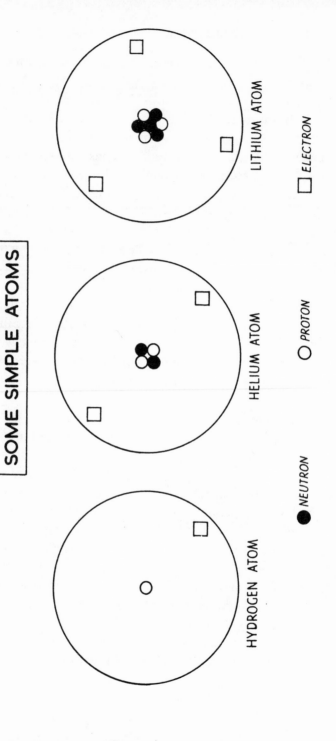

SOME SIMPLE ATOMS

HYDROGEN ATOM

HELIUM ATOM

LITHIUM ATOM

● NEUTRON

○ PROTON

□ ELECTRON

electron in each atom, will explode with the least encouragement. Helium, with two planetary electrons in each atom, won't react in any way no matter what. Consider the next element, the one with three planetary electrons. That is *lithium* (from a Greek word meaning "stone" because it was discovered in a mineral). It is not a gas, but is a solid under ordinary circumstances.

Three interesting elements are those with atomic numbers 6, 7, and 8. Element 6 is *carbon*. It is a solid, usually black material, which is very familiar to all of us since coal is a form of carbon. The word "carbon," in fact, comes from the Latin word for coal. Carbon is more important to living creatures than any other element.

Element 7 is *nitrogen*, and element 8 is *oxygen*. The atoms of nitrogen and oxygen pair off just as hydrogen atoms do. Nitrogen and oxygen are both gases. The air that is all around us is a mixture of the two, four-fifths nitrogen and one-fifth oxygen. The two gases differ in the way they behave. Nitrogen is rather inert (not as inert as helium, though), and oxygen is very active. When paper, wood, gasoline, hydrogen, and illuminating gas burn in air, it is the oxygen molecules with which they are reacting. If air consisted of pure oxygen and nothing else, these things would burn very brightly and rapidly indeed. If air consisted of pure nitrogen and nothing else, none of these things would burn at all. When you and I breathe, it is in order to draw the oxygen of the air into our lungs. The oxygen reacts with various molecules in our body and thus enables life to continue. The nitrogen is breathed in along with the oxygen, but it is merely breathed out again.

We shall not go through the list of elements, talking about each one. There isn't room for that. Instead, we shall list the names of some of the elements. I think you will be surprised at how familiar these names are to you:

Atomic Number	Element	Atomic Number	Element
10	Neon	50	Tin
13	Aluminum	53	Iodine
16	Sulfur	78	Platinum
24	Chromium	79	Gold
26	Iron	80	Mercury
28	Nickel	82	Lead
29	Copper	88	Radium
33	Arsenic	92	Uranium
47	Silver	94	Plutonium

There are elements, of course, whose names are completely unfamiliar to anyone but professional chemists. Element 41 is niobium, 49 is indium, 54 is xenon, 59 is praseodymium, 66 is dysprosium, 91 is protactinium, and so on. Fortunately, we won't have to bother much with them.

Families of Elements

Although each element is different from all the others, there are family resemblances among some of them. For instance, *sodium* and *potassium* resemble each other closely; both are soft, chemically active metals that are easily melted. *Bromine* is very much like *chlorine;* both are active, poisonous chemicals. *Argon* is very much like *neon;* both are inert gases.

The reason for such resemblances rests on the arrangement of the electrons of the atom. These electrons are not scattered about the outer regions of the atom in any old way. They are distributed in layers. It is as though there

SIMILAR ELEMENTS

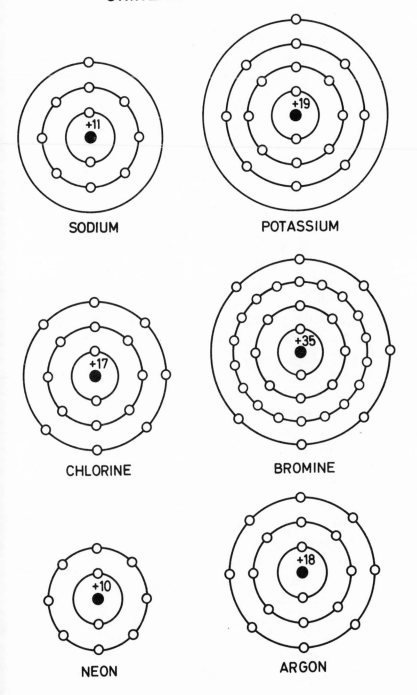

SODIUM

POTASSIUM

CHLORINE

BROMINE

NEON

ARGON

are *electron shells* surrounding the nucleus of each atom, like the layers in an onion. Each electron shell is capable of holding a particular number of electrons. The innermost can hold only two electrons, the next can hold eight, the next eighteen, and so on.

Let's consider the sodium atom now. It has an atomic number of 11, so that it has 11 electrons. These are distributed as follows: 2 in the innermost shell, 8 in the next, and 1 in the next. We can write that 2/8/1 and see that the total is 11.

If we turn to the potassium atom, with an atomic number of 19, we find that its electrons are divided up 2/8/8/1 for a total of 19. Knowledge concerning these electron arrangements came through the difficult theoretical labors of men like the Danish physicist, Niels Bohr, and the Austrian physicist, Wolfgang Pauli. For their work on the make-up of the electrons within the atom, both earned a Nobel Prize, Bohr in 1922 and Pauli in 1945.

If we compare the electron arrangements of sodium and potassium, 2/8/1 and 2/8/8/1, we see that they are alike in that both have a single electron in the outermost shell. When two atoms collide hard enough to undergo a chemical reaction, it is the outermost shell that gets the hard knock. If two atoms have similar outermost shells they will undergo similar kinds of reactions. The reactions won't be exactly alike because there are differences underneath, but they will be similar.

Sodium and potassium are similar because both contain a single electron in the outermost shell and they can be considered members of a family of elements that includes six altogether. One of these six is lithium, whose three electrons are arranged 2/1.

Chlorine and bromine are similar because of their electron arrangement, too. Chlorine, with an atomic number of 17, has its electrons arranged 2/8/7 (total, 17). Bromine, with an atomic number of 35, has its electrons arranged 2/8/18/7 (total, 35). Both have seven electrons in the outermost shell.

Then, there is the case of neon (atomic number 10) and argon (atomic number 18). The ten electrons of neon are arranged 2/8, while the eighteen electrons of argon are arranged 2/8/8. Both have eight electrons in the outermost shell.

All 105 elements can be divided into families, some of which contain as many as fifteen members, on the basis of their electron arrangement.

In fact, back in 1869, a Russian chemist, Dmitri I. Mendeléev, arranged the elements then known into a table of rows and columns. He knew nothing about electrons, and neither did anyone else in those days. He arranged the elements, however, in such a way as to make those with similar chemical properties fall into the same columns. This was called the *periodic table*.

For over fifty years chemists made use of the periodic table and found it very helpful in making sense out of chemical reactions. Then when they learned about electron arrangements, they found that Mendeléev, without knowing it, had arranged the elements in accordance with how many electron shells each possessed and how many electrons were in the outermost shell.

Charged Atoms

A few pages back, I talked about molecules colliding with one another. Sometimes, during these collisions, an atom in a molecule may lose electrons as a result, or perhaps pick up a few extra electrons. You may wonder if this change in the electron situation may not alter the atom itself and make it different from what it was. It does indeed.

Suppose we take sodium, which contains, of course, eleven electrons. When a sodium atom collides with another type of atom, the sodium atom frequently loses one of its electrons. It is left with only ten electrons.

What difference does that make? For one thing, as soon as an electron is lost, the sodium atom is no longer neutral. Originally, its eleven electrons were exactly balanced by eleven protons in the sodium nucleus. The atom was uncharged. After one electron is knocked off, the ten remaining electrons have a total charge of −10 while the eleven protons in the nucleus have a total charge of +11. The atom as a whole has a charge of +1.

Take another case. Element 17 is chlorine, a greenish, poisonous, active gas. Its atoms pair off the way hydrogen, oxygen, and nitrogen atoms do. When a chlorine molecule collides with another type of atom or molecule, the chlorine atoms within the molecule frequently pick up an extra electron apiece and keep them. The chlorine atom ends up with eighteen electrons, which have a total charge of −18. In the nucleus of a chlorine atom, however, there are only seventeen protons (that's not changed), with a total charge of +17. The atom as a whole has a charge of −1.

All atoms lose or gain one or more electrons if the conditions are right. Sometimes, as a result, molecules may have more or less than their usual number of electrons. Such atoms or molecules (or even, sometimes, atom-groups that are parts of molecules) are not neutral; they have positive charges (if electrons are missing) or negative charges (if they have extra electrons). The size of the charge depends upon the number of missing or extra electrons.

These charged atoms or molecules can move with an electric current, under the proper conditions, just as electrons do. Those with a negative charge move in the same direction as electrons. Those with a positive charge move in the opposite direction.

Because these charged atoms or molecules move in the presence of electric currents, they are called *ions*, from a Greek word meaning "to go." According to the type of charge, there are negative ions and positive ions.

When a positively-charged particle, such as a proton, flashes through matter, it can attract electrons away from the atoms it passes. A negatively-charged particle, such as a speeding electron, can repel electrons away from the atoms it passes. In either case, the atoms have fewer than the proper number of electrons and become positive ions. It is these ions that make it possible to track the speeding particles by means of water droplets in a cloud chamber or vapor bubbles in a bubble chamber.

An uncharged particle, such as a neutron, neither attracts electrons nor repels them. It lets them remain in place and therefore does not form ions. That is why neutral particles cannot be detected as easily as charged particles can be.

An electric current can be carried by ions present in water just as it can be carried by electrons present in a metal wire. The electricity in the storage batteries of automobiles is carried partly by ions.

Very pure water is not a good conductor of electricity, but it improves if ions get into it. Bath water, for instance, contains a number of ions derived from soap and from the salt in the body's perspiration. It is fairly good as a conductor. That is why it is dangerous to fiddle with electric equipment while in the bath. If there is a defect in the wiring, or if bare wires are exposed, electricity can be carried through the bath water to all parts of the body. Newspapers frequently tell of people who have been killed in this manner.

Now that we have been introduced to the various elements, it is time to take a still closer look at them. In the next chapter we will see if the different atoms in an element are really all alike.

ATOMIC TWINS

A Close Look at an Element

Of the elements listed a few pages back, one of the most familiar is *copper*. All of us have seen that red-orange metal. Electric wires are usually made of copper. One-cent pieces, or pennies, are 95 percent copper. Even five-cent pieces, or nickels, which are silvery in appearance, are 75 percent copper. (The other 25 percent is the element *nickel*, which gives the coin its color and its name.)

Now suppose we have a quantity of absolutely pure copper before us. By "pure" we mean that there is nothing in the copper except copper. No other metal, no other element at all, is present. What can we say about the atoms in that mass of copper?

In the first place, all the atoms have the same atomic

number. The atomic number of copper is 29, and that means that every single neutral copper atom must have 29 electrons. If an atom has more or less than 29 electrons, it is not a copper atom.

In addition, there must be 29 protons in the nucleus of each copper atom to balance the 29 electrons. Any atom that has more or less than 29 protons in the nucleus is not a copper atom.

That takes care of protons and electrons, but there is a third kind of sub-atomic particle which we mustn't forget. How many neutrons are there in each copper atom?

Here at last we find some variation. Some copper atoms contain 34 neutrons, and some contain 36 neutrons. To be exact, in any sample of copper that you can find, 69 percent of the atoms contain 34 neutrons, and 31 percent contain 36.

The number of neutrons present in the copper atom makes no difference in the atomic number. That depends only on the number of protons in the nucleus. It doesn't make any difference in the chemical behavior of the copper atoms. That depends only on the number and arrangement of the planetary electrons.

Does the number of neutrons make any difference at all, then? Well, the mass numbers of the two types of copper atoms are different. A 34-neutron copper atom has mass number 63 (29 protons plus 34 neutrons). A 36-neutron copper atom has mass number 65 (29 protons plus 36 neutrons).

The two types of copper atoms are identified by these mass numbers. When a chemist speaks of *copper-63,* he means a copper atom containing 34 neutrons in its nucleus. When he speaks of *copper-65,* he means one containing 36 neutrons in its nucleus.

Suppose you had a cubic inch of copper-63, containing

only atoms with 34 neutrons in their nuclei. How would it compare with a cubic inch of copper-65, containing only atoms with 36 neutrons in their nuclei? The two cubes would look the same. Both could be drawn into wires or pressed into pennies in the same way. The wires and pennies would behave exactly the same. If chemists treated them with acids or other chemicals, the two types of copper would act exactly alike. Copper-63 and copper-65 are atomic twins.

But suppose you weighed the two. The cubic inch of copper-63 would have a mass of 5⅛ ounces. The cubic inch of copper-65 would be more massive since each atom would contain two extra neutrons. All those additional neutrons would be enough to make the cubic inch of copper-65 have a mass of 5¼ ounces. There is only an eighth of an ounce difference, but that is enough to show that atomic twins aren't completely alike. (You'll find out, in future chapters, that there are other and more important differences between such atoms.)

Whenever two or more kinds of atoms differ from one another only in the number of neutrons in the nucleus, they are called *isotopes*. Copper-63 and copper-65 are copper isotopes.

Isotopes Many and Isotopes Few

You may ask why copper atoms have either 34 or 36 neutrons in the nucleus. Why not 35? Why not 37? Why not any number except 34 or 36?

Well, let's see. You may remember that when we first talked about positive and negative electricity, we said that like charges repel one another. Protons all carry positive charges, so two protons ought to repel each other — and they do!

Yet 29 protons can be squeezed together into the tiny

nucleus of the copper atom and stay together. What keeps them from repelling one another and flying apart? Apparently part of the answer is the presence of neutrons. Whenever a nucleus contains more than one proton, it must also contain neutrons. What's more, every combination of protons requires at least a certain number of neutrons.

In the case of the copper atom, 34 neutrons distributed among the 29 protons of the nucleus will keep that nucleus from flying apart. So will 36 neutrons. A copper atom containing either 34 or 36 neutrons in its nucleus stays put. It is *stable*. Any number of neutrons other than 34 or 36 fails to keep the protons together. If there were 35 neutrons, for instance, the nucleus would not stay put. A copper atom containing 35 neutrons (or any number other than 34 or 36) in its nucleus is *unstable*.

Copper is therefore said to have two stable isotopes.

Some elements can have more than two stable isotopes. The element *iron* is a good example. The atomic number of iron is 26. All neutral iron atoms contain 26 electrons and 26 protons. Different iron atoms differ, however, in the number of neutrons they contain in their nuclei. Fully 92 percent of all iron atoms contain exactly 30 neutrons in their nuclei. These are atoms of iron-56. (You can see that 26 protons and 30 neutrons come to mass number 56.) The remaining 8 percent, however, include three different varieties of iron atoms. There is a kind with only 28 neutrons in its nuclei (iron-54). Another kind has 31 neutrons (iron-57), and still another kind has 32 (iron-58). Each type of atom is stable.

Iron, therefore, consists of four stable isotopes.

In number of stable isotopes the champion element is *tin*. It actually has ten stable isotopes. Chemists can speak of tin-112, tin-114, tin-115, tin-116, tin-117, tin-118, tin-119, tin-120, tin-122, and tin-124. Any piece of tin you come across will contain a little of each isotope.

If we look closely at the different elements and their isotopes, we may see an interesting point. Elements with an even atomic number (that is, with an even number of protons in the nucleus) have more isotopes than elements with an odd atomic number.

Protons are easier to handle, apparently, if an even number of them occur in an atomic nucleus. They are paired off, and the nucleus then seems to be better balanced. At least, it is easier to stabilize such a nucleus by adding neutrons. The exact number doesn't seem to be very important. Most of the elements with even atomic numbers have three or more stable isotopes, which means that any of three or more different numbers of neutrons will do the trick.

Iron (atomic number 26) has four isotopes, as we have said. So has *chromium* (atomic number 24). *Nickel* (atomic number 28) has five, and tin (atomic number 50) has, as we just mentioned, ten. Fifty protons in a nucleus can be stabilized by 62 neutrons, or 64, or 65, or 66, or 67, or 68, or 69, or 70, or 72, or 74.

On the other hand, it is quite tricky to balance a nucleus containing an odd number of protons. After as many protons as possible have paired off, there is an odd proton left over. The odd proton makes the nucleus unbalanced or rickety.

No element with an odd atomic number has more than two stable isotopes. In other words, for any such element there are, at the most, only two numbers of neutrons that will satisfy the protons and keep them from breaking up their association. Copper (atomic number 29) is one example, which we have already mentioned. Another is *silver* (atomic number 47). The two stable isotopes of silver are silver-107 and silver-109.

Most of the elements with odd atomic numbers are even more particular than that. They have only one stable variety of atom apiece. An example is *aluminum*. Its atomic number is 13. The nucleus of its atom, to be stable, must contain 14 neutrons, no more and no less. No other number will do. The only stable atom variety of aluminum is therefore aluminum-27.

Similarly, the only stable kind of *arsenic* (atomic number 33) is arsenic-75. The only stable kind of *iodine* (atomic number 53) is iodine-127. The only stable kind of *gold* (atomic number 79) is gold-197. And so it goes.

Sometimes, an element with only one stable kind of atom is said to have "only one isotope." Actually, though, the term, isotope, should only be used where an element has two or more varieties. Each single variety is called a *nuclide*, so you can say that aluminum, for instance, consists of a single nuclide.

There are even some elements that have no stable atoms of any sort at all. Actually none! Not one! (This may puzzle you and raise several questions in your mind, but be patient. We'll talk a good deal about this later on.)

Of the 103 known elements, 23 possess no stable isotopes, 20 possess one stable isotope, and 60 possess two or more stable isotopes. Altogether there are 266 different stable isotopes known.

It is possible for two different elements to have isotopes of the same mass number. For instance, all *calcium* atoms have 20 protons in their nuclei and all *argon* atoms have 18 protons. But some calcium atoms contain 20 neutrons as well, while some argon atoms contain 22. The calcium atoms that contain 20 protons and 20 neutrons make up the isotope, calcium-40. The argon atoms with 18 protons and 22 neutrons are argon-40. Two atoms which have the same mass number but different atomic numbers are called *isobars.* Calcium-40 and argon-40 are examples of isobars.

Heavy Hydrogen and Heavy Water

The simplest atom that can exist is that of hydrogen. Its atomic number is 1. It has only a single electron. Its nucleus contains a single proton and, in the commonest kind of hydrogen atom, nothing else, no neutrons at all. Such a hydrogen atom has mass number 1 and is therefore called hydrogen-1.

Does hydrogen possess any other isotopes? The answer is yes. An atomic nucleus containing one proton and one neutron is also stable. An atom possessing such a nucleus still has atomic number 1 (since there is still only one proton in the nucleus) and still has only one electron. Such an atom is still hydrogen. Its mass number, however, is 2 (one proton plus one neutron). It is therefore called hydrogen-2.

Hydrogen-2 occurs in ordinary hydrogen gas in very small amounts. For every atom of hydrogen-2 in hydrogen, there are five thousand atoms of hydrogen-1.

The interesting thing about the hydrogen isotopes is that they are more different in mass than the isotopes of any other element. This may surprise you. You may say that the hydrogen isotopes differ in mass by only one unit. The two copper isotopes, copper-63 and copper-65, differ by two units. Two of the tin isotopes, tin-112 and tin-124, differ by twelve units. Why are the hydrogen isotopes so special then?

You see, a small difference among small numbers is much more important than that same difference (or even a slightly larger difference) among large numbers.

Suppose that you showed a youngster two heaps of marbles, in one of which were 63 marbles and in the other 65. Without telling him how many marbles are in each heap, you ask him to tell you which heap is larger. The question would probably puzzle him. He would have to guess, or, if he were completely frank, he would say that they looked about the same to him. The situation would be the same if one heap contained 112 marbles and the other 124.

That is the sort of situation we have with the copper and tin isotopes. Copper-63 or copper-65. Tin-112 or tin-124. The mass numbers are so large that a difference of two, or even twelve, isn't much of a difference.

Now suppose the same youngster is faced again with two heaps of marbles, one containing a single marble and one containing two marbles. Ask him again which heap contains more marbles. A glance is sufficient this time. The difference between 1 and 2 is much more obvious than the difference between 63 and 65 or between 112 and 124.

In short, the difference between hydrogen-1 and hydrogen-2 is only a single unit, but the numbers are so small that that single unit is enough to make hydrogen-2 twice as massive as hydrogen-1.

Another way of looking at it is this: We've already said that a cubic inch of copper-63 has a mass of 5⅛ ounces, while a cubic inch of copper-65 has a mass of 5¼ ounces. There is an eighth of an ounce difference. If, instead, we took 5⅛ ounces of tin-112 and changed each atom to tin-124, the mass would become 5⅘ ounces. There would then be about ⅔ of an ounce difference.

But now suppose we take 5⅛ ounces of hydrogen-1 and

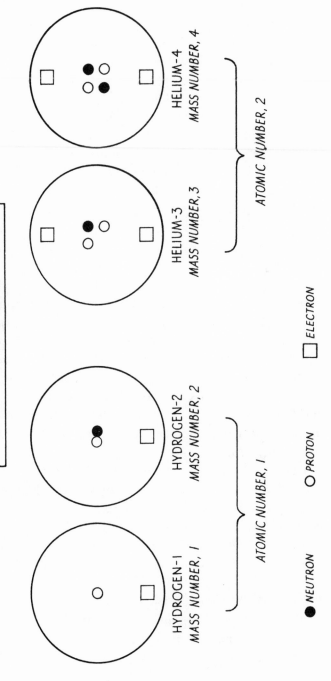

SOME SIMPLE ISOTOPES

HYDROGEN-1
MASS NUMBER, 1

HYDROGEN-2
MASS NUMBER, 2

ATOMIC NUMBER, 1

HELIUM-3
MASS NUMBER, 3

HELIUM-4
MASS NUMBER, 4

ATOMIC NUMBER, 2

● NEUTRON   ○ PROTON   □ ELECTRON

convert every atom to hydrogen-2. The mass becomes 10¼. It is exactly doubled. The difference in mass is not a fraction of an ounce, but a full 5⅛ ounces.

The greater mass of hydrogen-2 compared with that of hydrogen-1 is so noticeable that hydrogen-2 is commonly referred to as *heavy hydrogen.* The two hydrogens, in fact, are so different in mass that they have even been given separate names. Hydrogen-1 is sometimes called *protium* (from a Greek word meaning "first"), and hydrogen-2 is called *deuterium* (from a Greek word meaning "second").

The chemical behavior of the two hydrogen isotopes depends only on the number of their electrons. Since both possess only one electron, they behave very much alike. Despite their difference in mass, they are still atomic twins. Both, for instance, will combine with an oxygen atom to form a water molecule. (The atoms in a molecule are sometimes alike, as in the hydrogen molecule or the oxygen molecule. Usually, though, a molecule contains atoms of different kinds. The water molecule, for example, contains two hydrogen atoms and one oxygen atom.) The result is that there are water molecules containing two atoms of hydrogen-1, a few which contain one atom of hydrogen-1 and one atom of hydrogen-2, and a very few which contain two atoms of hydrogen-2.

The water molecules which contain hydrogen-2 are called *heavy water* because they are more massive than the ordinary variety. Heavy water is very rare: only one molecule of water in twenty-five million contains two atoms of hydrogen-2. There is so much water on the earth, however, that even this small fraction mounts up to a considerable total.

How Isotopes Were Discovered

We can talk freely about isotopes now, but it took people a long time to find out they exist.

It is fairly easy to tell elements apart. Even in prehistoric days, people knew the difference between some of them. Primitive metal-workers, for instance, weren't very likely to confuse copper and gold.

If you had a sample of copper and one of gold before you, you wouldn't confuse them either. Copper is reddish, and gold is yellow. Gold is much more massive than copper. A cubic inch of copper weighs a little over five ounces, but a cubic inch of gold weighs about eleven ounces. If a chemical called nitric acid were placed on a sheet of copper, bubbles would form on that spot, and both copper and acid would turn green. If a drop of the same chemical were placed on a sheet of gold, nothing at all would happen. There are many other differences, too.

All elements differ from one another in appearance and behavior. Usually the difference is even more noticeable than in the case of copper and gold. Chemists can take objects which contain a dozen or more different kinds of atoms and, by using various chemicals and instruments, can identify each element present and tell the quantity of each. This procedure is known as *chemical analysis*.

The problem of isotopes is a much more difficult one. Different isotopes of the same element are so similar in appearance and behavior that the ordinary methods of chemical analysis can't tell them apart.

Chemists first began to suspect that isotopes might exist in connection with the odd behavior of certain elements with particularly complicated atoms. I will begin discussing these elements in the next chapter.

The final clincher, though, came in 1919, when two different isotopes of a single element were actually separated by J. J. Thomson, the English scientist who, twenty-three years earlier, had discovered the electron.

He was working with the element *neon,* a rare, inert gas

somewhat similar to helium. It occurs in small amounts in the air and is used in the "neon lights" with which we are all familiar.

This is the way the isotopes of neon were discovered: An electric current was allowed to pass through a curved glass tube containing a small quantity of neon and nothing else. An electric current, as you know, is really just a moving stream of charged particles, usually electrons. The moving electrons often hit the neon atoms as they moved through the tube. Occasionally the electrons would strike an atom hard enough to knock an electron out of it. When that happened, the neon atom, minus one electron, became a positively charged ion.

The neon ion, like any charged particle, moves under the influence of the electric current. Ordinarily it would move in a straight line. However, the glass tube is surrounded by a magnet which bends the path of the positive ions and makes them follow the curve of the tube. At the far end of the tube is a piece of photographic film, which the moving ions finally strike. When the film is developed, it shows a dark spot where the ions have struck.

If all the neon ions were identical, they would all hit the same place on the film, and there would be only one spot on the developed negative. But the neon ions are not identical. Some of them are more massive than others. The more massive ions curve more slowly under the influence of the magnet. (You remember the difference between kicking a moving billiard ball and kicking a moving cannon ball, which we talked about in the first chapter.) As a result, the more massive ions and the less massive ions hit the photographic film in two different places. There are two dark spots on the film after it is developed.

Such an instrument, which separates ions of different

mass and therefore reveals the presence of different isotopes, is called a *mass spectrograph*. Thomson's first instrument was quite crude. One of his assistants, however, the English chemist, Francis William Aston, improved it and made it capable of separating ions very finely and accurately. Aston received a Nobel Prize in 1922 for this.

It is possible to calculate the mass of different ions from the exact positions of the dark spots on the negative. In that way it was found out that the neon isotopes are neon-20. and neon-22. Furthermore, the size of a particular dark spot depends on how many ions have struck. Scientists can therefore tell which isotope is present in greater quantity and by how much.

The spot made by neon-22, for example, is much fainter than that made by neon-20. From this, it can be shown that 90.5 percent of neon atoms are neon-20 and that only 9.2 percent are neon-22. (The remaining 0.3 percent are a third stable isotope, neon-21.)

All the other elements have been studied in this way or in similar ways since 1919, and their isotope composition has been determined.

How Isotopes Are Separated

A mass spectrograph separates isotopes into little heaps on the photographic film. The heaps are so small they are useless except to make marks on the film. Scientsts who wished to study pure isotopes were eager to find a way of separating larger quantities.

Fortunately, the differences in mass among the isotopes not only make the mass spectrograph possible, but also make large-scale separation possible.

Different isotopes of the same element, as we have seen, behave the same way in a particular set of circumstances.

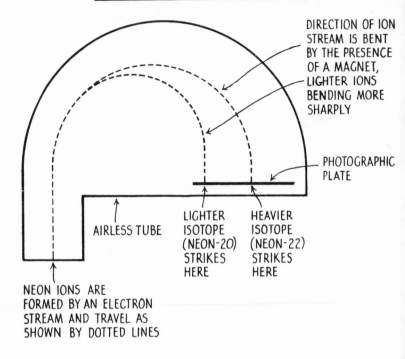

THE MASS SPECTROGRAPH

DIRECTION OF ION STREAM IS BENT BY THE PRESENCE OF A MAGNET, LIGHTER IONS BENDING MORE SHARPLY

PHOTOGRAPHIC PLATE

AIRLESS TUBE

LIGHTER ISOTOPE (NEON-20) STRIKES HERE

HEAVIER ISOTOPE (NEON-22) STRIKES HERE

NEON IONS ARE FORMED BY AN ELECTRON STREAM AND TRAVEL AS SHOWN BY DOTTED LINES

But (and this is important) the more massive isotopes move a little more slowly than the lighter ones. The more massive isotopes go through the same motions as the lighter ones, but always lagging a little behind. You yourself would run more slowly if you were carrying a ten-pound weight on your shoulder than if you were not. It is the same with isotopes. An additional neutron or two slows them up.

This difference between isotopes is useful to us in a number of ways.

Let us consider water. The water molecule, we know, contains two hydrogen atoms and one oxygen atom. If an electric current is passed through water, the water molecules break up. Hydrogen and oxygen (which are gases) are liberated and bubble up through the water.

But some of the water molecules contain hydrogen-1 and some contain hydrogen-2. Hydrogen-1, being lighter and more nimble than hydrogen-2, takes advantage of the electric current a little more quickly. The molecules containing hydrogen-2 lag behind and break up more slowly.

Suppose, then, that a large quantity of water, hundreds of gallons, is subjected to an electric current until almost all of it has been broken up into hydrogen and oxygen. What would be left would be almost entirely water containing hydrogen-2, or heavy water, as it is called.

You can see why this is so if you imagine a long race in which many contestants are entered. At the start, all the contestants, the faster ones and the slower ones, are jumbled together. By the time the end of the race approaches, though, all the slow runners have collected at the rear of the group. You have separated the fast runners from the slow runners.

Another way to separate heavy water from ordinary water is to boil a large quantity of water slowly. The molecules containing hydrogen-2 are a little slower to boil off than the molecules containing hydrogen-1. Again, the last few drops are almost pure heavy water.

By either of these methods (and others, too) large quantities of heavy water have been prepared. Such heavy water can be treated with an electric current, its molecules broken up, and pure heavy hydrogen collected.

The first man to succeed in obtaining heavy water and to show that hydrogen-2 really existed was the American chemist, Harold Clayton Urey. His success came in 1931, and he received a Nobel Prize in 1934 in consequence.

By various methods, always taking advantage of the sluggishness of the more massive isotopes, the isotopes of other elements have also been separated.

ATOMIC BREAKDOWNS

Unstable Atoms

In the previous chapter we said that, if protons are to remain together inside an atomic nucleus, neutrons also must be present. Let's consider that a bit more.

At first, when the number of protons is small, an equal number of neutrons is enough to keep the nucleus stable. The nucleus of helium-4 contains 2 protons and 2 neutrons. The nucleus of carbon-12 contains 6 protons and 6 neutrons. The nucleus of oxygen-16 contains 8 protons and 8 neutrons. The nucleus of neon-20 contains 10 protons and 10 neutrons.

This state of affairs doesn't continue for very long, however. When an atomic nucleus contains more than 20 protons, an equal number of neutrons is no longer enough to make the nucleus stable. The job of stabilization seems to grow harder as the protons pile up, and extra neutrons must be added.

The iron atom, for instance, has 26 protons in its nucleus, but 26 neutrons are not enough to make a stable nucleus for iron. It takes 28 at least. That's 2 extra neutrons. The copper atom, containing 29 protons in its nucleus, requires at least 34 neutrons to be stable, and that's 5 extra. The tin atom, containing 50 protons in its nucleus, requires at least 62 neutrons to be stable, and that's 12 extra.

The situation keeps getting worse. The most massive stable atoms are varieties of *lead* (atomic number 82) and *bismuth* (atomic number 83). Bismuth has a single isotope, bismuth-209, with 83 protons and 126 neutrons in the nucleus. There are 43 more neutrons than protons there. Lead is made up of four stable isotopes, of which the most massive is lead-209. This contains 82 protons and 126 neutrons in its nucleus, or 44 more neutrons than protons.

When the number of protons in an atomic nucleus is greater than 83, the whole system breaks down. An atomic nucleus with more than 83 protons can never be stable, apparently, no matter how many neutrons are put on the job.

Yet atoms containing more than 83 protons in the nucleus do exist. A fairly common type of atom in the earth's soil is *uranium*. Uranium has atomic number 92, and its most common isotope is uranium-238. Uranium-238 contains 92 protons and 146 neutrons in its nucleus. (That's 54 extra neutrons.)

Despite all those neutrons, uranium-238 is not stable. Yet there it is. It exists.

To explain this seeming contradiction, let's compare

uranium atoms to human beings. All human beings are mortal. That is, each human being will die some day. However, that doesn't mean an individual can't live for years before dying. A particular human being may die this minute, true. Another, on the other hand, may not die for a hundred years.

If you were to take a million newborn Americans at random and follow their histories, you would find that ten or twenty of them might die each day. After sixty-five years or so, nevertheless, half of them might still be alive.

It is the same with uranium atoms. They are unstable, but they don't all break down at once. One atom may break down this minute, true. Another may last for a hundred years. Still another may last for many billions of years. Since the beginning of our planet, millions and billions of uranium atoms have been breaking down every second. They are still breaking down this very second. Nevertheless, most of the original atoms still exist and will not break down till some time in the future.

You may think: If some uranium atoms don't break down for billions of years, that seems pretty stable. Why are they called unstable?

There's a great difference between a billion years and forever. A really stable atom, such as one of oxygen-16, never breaks down at all.

A man may live a hundred years or even more and yet not be immortal. You might say that an oxygen-16 atom is immortal if it is left alone, but a uranium-238 atom is not.

Another thing to remember is that when an unstable nucleus breaks down, it does not explode like dynamite. The process is much more orderly. The unstable nucleus simply throws out a sub-atomic particle or two. In any mass of uranium atoms, a certain number are breaking down each moment and spraying, or "radiating," these sub-atomic particles in every direction.

This behavior of uranium was first discovered in 1896 by a French physicist, Antoine Henri Becquerel. It seems that Becquerel's father (also a physicist) had been particularly interested in certain minerals which gave off glowing colors when exposed to sunlight. (This is called *fluorescence*.) One of the minerals Becquerel's father studied contained uranium atoms in its molecule, and Becquerel thought he would study the glow of that particular mineral.

He was chiefly interested to see if that colored glow was powerful enough to pass through paper, for certain forms of radiation which had just been discovered could do so. Consequently, Becquerel placed a piece of photographic film in the sunlight after he had covered it with black paper. The sunlight didn't penetrate the black paper, and the photographic film was left untouched. Next he put some of the uranium mineral on the paper. This time, the film was fogged. It seemed that the colored glow of the uranium could penetrate the paper.

Becquerel wanted to continue the experiment, but a long siege of cloudy weather hit Paris. There was his film, with the black paper on top of it and the mineral on top of that — but no sunlight. Finally, just to keep busy, he developed the film and found it was fogged very strongly.

Something from the uranium mineral was penetrating the paper even when there was no sunlight. It had nothing to do with either sunlight or the colored glow. The uranium mineral was giving off strong, penetrating radiations at all times. What's more, it quickly turned out that the radiations were coming from the uranium atom.

Since uranium was so actively giving off these radiations,

the phenomenon was called *radioactivity* by a Polish-born French chemist, Marie Sklodowska Curie, who had grown interested in the new discovery. Uranium was the first *radioactive element* discovered, but others followed. In 1898, Madame Curie found that the element, *thorium* (atomic number 90), was radioactive.

In 1903, Becquerel received a Nobel Prize for his discovery. Sharing it with him were Madame Curie and her husband, for work which I will shortly tell you about.

The Three Radiations

Before very long, scientists discovered that the radiations of uranium are of three different varieties.

When they exposed the radiations to the influence of a magnet, they observed three kinds of behavior. The path of one kind of radiation was bent slightly by the influence of the magnet. (They could tell this by noting what part of a photographic film was hit by the beam of radiation, and in other ways, too.) The path of a second kind was bent a great deal in the direction opposite to that in which the first kind was bent. A third kind of radiation wasn't affected by the magnet at all.

The first radiations were named *alpha rays* by Rutherford, the discoverer of the atomic nucleus. The second were named *beta rays*. The third were named *gamma rays*. The words "alpha," "beta," and "gamma" are simply the names of the first three letters of the Greek alphabet.

Eventually the alpha rays and beta rays were found to consist of sub-atomic particles. These days, therefore, it is usual to speak of these radiations as streams of *alpha particles* and *beta particles.*

The beta particle was identified in 1900 by Becquerel. It was found to be a rapidly moving electron.

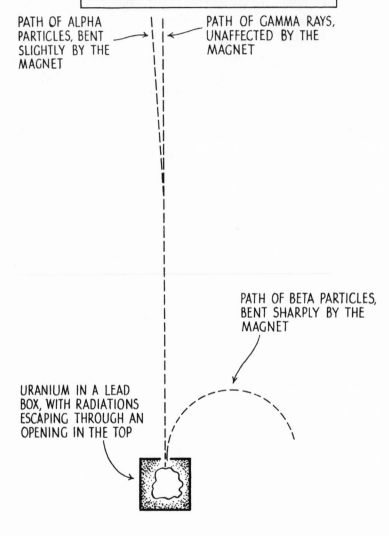

RADIOACTIVE RADIATIONS
IN THE PRESENCE OF A MAGNET

PATH OF ALPHA PARTICLES, BENT SLIGHTLY BY THE MAGNET

PATH OF GAMMA RAYS, UNAFFECTED BY THE MAGNET

PATH OF BETA PARTICLES, BENT SHARPLY BY THE MAGNET

URANIUM IN A LEAD BOX, WITH RADIATIONS ESCAPING THROUGH AN OPENING IN THE TOP

Because of this, physicists concluded that there must be electrons in the atomic nucleus. For twenty years after Rutherford discovered the nucleus, attempts were made to work out its structure in terms of protons and electrons only.

These created certain puzzling difficulties. These were cleared up, eventually, when Chadwick discovered the neutron and Heisenberg pointed out the nucleus must be made up of protons and neutrons only. By 1933, it seemed quite clear there were no electrons in the nucleus.

But in that case where do the beta particles come from? How can they be speeding electrons and come shooting out of the nucleus if there are no electrons in the nucleus in the first place?

What happens inside the nucleus when a beta particle is thrown out is this: A neutron is changed to a proton.

Suppose a neutron were actually a proton and an electron mashed together into one particle. It would be uncharged, of course, since the positive charge of the proton would cancel the negative charge of the electron. If the neutron were somehow to get rid of the electron it contained, and send it out of the nucleus in the form of a beta particle, a positive charge would be left. The neutron would have become a proton.

Neutrons, when free, break down just as they sometimes do in radioactive nuclei. A stream of neutrons is continually breaking down to form protons and electrons so that an individual neutron only lasts a quarter of an hour on the average. Even in stable nuclei, the neutrons are probably breaking down, changing to protons; while the protons in the nuclei change to neutrons. This rapid changing back and forth helps keep the nucleus stable.

It took several more years to identify the alpha particle.

From the direction in which a magnet bent streams of alpha particles, it could be seen that the particles must carry a positive charge. The bending was so slight, moreover, that alpha particles had to be even more massive than protons.

It turned out, eventually, that alpha particles have mass number 4 and a charge of $+2$. In other words, they consist of two protons and two neutrons packed tightly together, just as in the nucleus of helium-4. Alpha particles, then, are rapidly moving helium nuclei.

The final proof of this was obtained by Rutherford in 1909. He trapped some alpha particles. As they slowed down, they picked up some electrons from their surroundings and became ordinary atoms of helium. Rutherford was able to identify the gas that formed as helium.

This combination of two protons and two neutrons is a particularly stable one. Inside the atomic nucleus, protons and neutrons seem to group themselves into such combinations of four. Apparently, when an unstable nucleus breaks down, it often converts itself into something simpler by getting rid of one of these stable two-proton-two-neutron combinations all in one piece. In this way, alpha particles are produced.

The gamma rays are something entirely different. To consider them properly, we must first change the subject temporarily.

The Different Kinds of Light

All of us are familiar with *light*. We are so familiar with light that we take it for granted and probably never stop to think about it. But what is light, after all?

It doesn't weigh anything. It doesn't have mass. It takes up no room. (You can fill a room with brilliant light, but

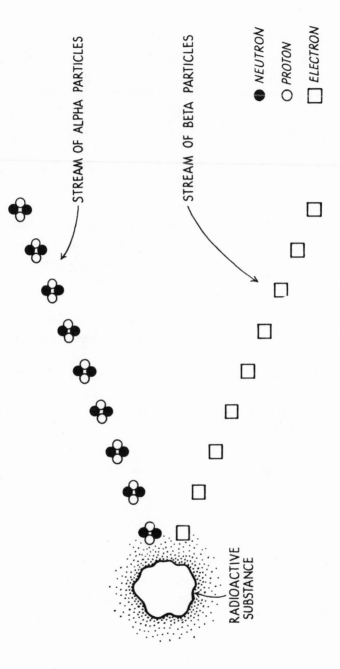

ALPHA PARTICLES AND BETA PARTICLES

STREAM OF ALPHA PARTICLES

STREAM OF BETA PARTICLES

● NEUTRON

○ PROTON

□ ELECTRON

RADIOACTIVE
SUBSTANCE

nothing has to get out of the way. There is still as much room for all the air and furniture and people in the light as in the dark.)

Light is not matter at all. Light is a form of *energy*. It is not easy to explain exactly what energy is, but we can give examples. The most widespread and simplest form of energy is *heat*. We are all familiar with that, too, and know what it feels like. The feeling we call heat is the result of the aimless motion or vibration of molecules. (Such aimless motion in every direction is called *random motion.*) The faster the random motion of the molecules in an object, the hotter that object is. Our bodies have a type of nerve ending in the skin that can detect such molecular motion. We know if an object is cold or warm or hot, or if one object is warmer than another. Scientists have developed instruments that can measure heat more delicately than the human skin can.

Anything which can be converted into heat—that is, into random molecular motion—is also a form of energy. Electricity can be converted into heat (feel a burning light bulb quickly if you want proof), and so can magnetism, and so can sound. All are forms of energy. Objects moving in one direction (this is *directed motion*) contain energy. If the directed motion is stopped, it is converted into random molecular motion, which is heat. (If you pound nails long enough, you will notice that the head of the hammer grows warm.) All substances contain a certain amount of chemical energy. This is most noticeable in substances which burn easily and throw off heat (and light). And light, of course, is energy, too.

In 1900, the German physicist, Max K. E. L. Planck, showed that the energy of light could be considered as divided up into little pieces called *quanta* (singular, *quantum*). He received a Nobel Prize in 1918 for this. The importance of the discovery was increased when it quickly turned out that all forms of energy could be divided into such quanta, just as matter is divided up into electrons, protons, and neutrons.

Quanta are extremely small pieces of energy just as sub-atomic particles are extremely small pieces of matter. However, protons, neutrons, and electrons are the same in all matter, whereas quanta can be very different in size, depending on the type of energy.

Red light, for instance, is made up of rather small quanta. The quanta of orange light are a bit larger, those of yellow, green, and blue light larger still in that order. The quanta of violet light are larger than any of these. They are twice the size of the quanta of red light. The reason why we see colors, in fact, is that quanta of different size have different effects on our eyes.

The larger the quanta the more "energetic" the particular kind of light. This was seen to be so when physicists studied the manner in which light, shining upon the surface of certain metals, caused electrons to go shooting out of the metal. This is called the *photoelectric effect*.

The German physicist, Philipp Eduard Anton Lenard, showed in 1902 that violet light whipped out electrons at much greater velocities than red light did. Sometimes red light couldn't force any electrons out of the metal at all.

In 1905, another German physicist, Albert Einstein, used Planck's quantum theory to explain this. Since Planck had advanced it five years earlier, little had happened because not many scientists could believe in the existence of quanta. Once Einstein showed that the electrons came out of the metal with a velocity that increased as the size of the light-quanta increased, all that was changed. The quantum theory came to be accepted.

As a result Lenard received a Nobel Prize in 1905, and Einstein received one in 1921.

Large light-quanta bring about chemical changes more easily than small ones do. That means blue and violet light can easily bring about certain chemical changes that red light can't bring about at all. We can see this in the case of the chemical change that light brings about in photographic film. (That is what accounts for the black and white pattern we call a photograph after the film is developed.) In most ordinary types of film, only the more energetic forms of light can do the trick. The quanta of red light are too small to affect the film. That is why darkrooms in which photographic films are being developed are sometimes lit by a small red bulb. The photographer can see by the red light, but it will not ruin the film.

Is there any form of light with quanta still smaller than those of red light? The answer is yes. Such light is called *infrared*. (The word "infra" is Latin and means "below.") Infrared light doesn't affect our eyes at all, so it cannot be seen. If it strikes our skin, it is converted into heat, and we can feel it as such. There are special "heat lamps" with which some of you may be familiar. When they are on, they glow a deep red, but most of the light they give off is the invisible infrared. People sometimes use such lamps to ease the pain of aching muscles.

The existence of infrared light was first demonstrated in 1800 by a German-English astronomer, William Herschel. In 1870, however, a Scottish mathematician, James Clerk Maxwell, predicted that this was only the beginning. He worked out a theory that led him to believe there was a whole stretch of lightlike radiations.

In those days, it was known that light existed as very tiny waves. The different colors of light differed in the size of these waves. The wavelength of light can be measured in *millimicrons*, where one millimicron is equal to only 1/25,000,000 of an inch. Thus, violet light has the very short wavelength of about 350 millimicrons, whereas red light has wavelengths as high as 700 millimicrons. The average light wave is thus about 1/50,000 of an inch in length.

The shorter the wavelength, the larger the quanta. The short-wave violet light has quanta that are twice as large as those of the long-wave red light. The infrared light, which has wavelengths still longer than those of visible red light, has quanta even smaller.

Maxwell's prediction was that there would be forms of light with wavelengths far longer even than that of infra-red light. He also predicted forms of light with wavelengths far shorter than that of violet light.

In 1888, a German physicist, Heinrich Rudolf Hertz, showed that Maxwell was correct. He found evidence for invisible radiations with wavelengths much longer than those of the infrared. Because these waves have come to be used in transmitting messages by radio they are called *radio waves*. Particularly short radio waves, used in radar, for instance, are called *microwaves*. Microwaves are made up of quanta that are smaller than those of infrared, and the quanta of radio waves are smaller still.

Now let's work in the other direction. What about light with wavelengths shorter than those of violet light, and, therefore, with quanta that are bigger?

Back in 1801, a German physicist, Johann Wilhelm Ritter, had discovered invisible radiations that could bring about chemical changes even more easily than violet light could.

These radiations had wavelengths shorter than those of violet light and therefore came to be called *ultraviolet light*. (The word "ultra" is Latin and means "beyond.")

We can't see ultraviolet, but its quanta are large enough to damage the eye. People who work with ultraviolet light must always be careful to protect their eyes by special goggles. Ultraviolet is also energetic enough to harm the skin. After exposure to ultraviolet light the skin will redden, and after too much exposure it will blister. It is the ultraviolet light in sunshine that causes sunburn. Ultraviolet light will affect a photographic film, of course, and so will forms of light that are still more energetic.

Maxwell's theories, however, predicted forms of radiation with wavelength much shorter even than ultraviolet. Did these exist?

Evidence for such radiation arose from work with cathode rays (see page 19). In 1895, a German physicist, Wilhelm Konrad Roentgen, was experimenting with cathode rays. He found that when a cathode ray was in operation, some radiation was making a chemical-coated paper in his laboratory fluoresce, so that it shone in the dark.

He turned his attention to this fact and found that the radiation could penetrate paper and wood. It could even penetrate thin sheets of metal. The discovery was so astonishing that Roentgen was awarded a Nobel Prize in 1901, the first year in which these prizes were distributed.

Roentgen called the radiation *x-rays*. (The letter "x" is often used in mathematics and science to stand for something unknown.) Later on, it was discovered that x-rays are a type of light composed of quanta still larger and wavelengths still shorter than those of ultraviolet. Sometimes the radiation is known as *Roentgen rays* in honor of its discoverer, but "x-rays" is still the common name, even though it is no longer mysterious.

X-rays are so energetic that they can shoulder their way right past atoms, particularly those with low mass numbers. For instance, ordinary human flesh consists largely of hydrogen (mass number 1), carbon (mass number 12), nitrogen (mass number 14), and oxygen (mass number 16). X-rays pass through flesh without much trouble. Bones and teeth, however, contain a great deal of phosphorus (mass number 31) and calcium (mass number 40). X-rays have difficulty getting past those.

Doctors and dentists sometimes aim a beam of x-rays at a photographic film and place a part of the human body in the path of the beam. The film, after being developed, shows white where x-rays were stopped, gray where they passed through with difficulty, and black where they passed through easily. Information about the body can be obtained from this pattern by people who are experienced in such things. You have probably all had your teeth and chest x-rayed at some time or another, so you know what that's like.

Unfortunately, x-rays are even more energetic than ultraviolet and can be even more dangerous. They must be used only in small quantities and never without a doctor's direction.

Gamma Rays

The less energetic forms of the group of radiations we have been talking about — radio waves, microwaves and infrared light — are produced by certain types of vibrations of atoms and molecules. The more energetic forms are produced by energy shifts of electrons within the atoms. The closer the particular electrons are to the atomic nucleus,

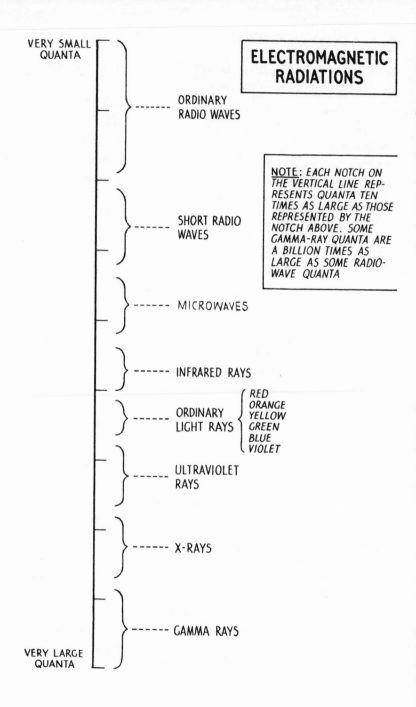

the larger the quanta produced by them. Energy shifts of the outermost electrons of an atom produce the different types of visible light. Energy shifts of electrons further in produce ultraviolet light. Energy shifts of the innermost electrons produce x-rays.

But what about energy shifts of the particles that are inside the nucleus? That should produce still larger quanta, larger than anything we've mentioned. They do! When a nucleus throws out an alpha particle or a beta particle, the particles that are left in the nucleus often rearrange themselves into the most stable new position. In this rearrangement an extremely large quantum, even larger than that of x-rays, is given off. This extremely energetic radiation is what we call gamma rays.

Gamma rays penetrate matter even more easily than x-rays do. They are even more dangerous.

This entire range of radiations, from radio waves down to gamma rays, are called *electromagnetic radiations* because they can be produced by particles carrying an electric charge and producing a magnetic field. All electromagnetic radiations, of whatever type, travel at the same speed. This is about 186,282 miles per second in a vacuum (and just very slightly slower in air). This extremely high speed, sometimes called the "speed of light," makes it possible for a man to speak in Los Angeles and be seen and heard almost at once on a television set in New York. The radiations picked up by the television set cross the United States in 1/60 of a second.

Quanta, particularly the more energetic quanta, behave something like particles. The American physicist, Arthur Holly Compton, proved this in 1923, when he showed that x-rays could push electrons out of the way just as though the x-rays were themselves particles. Compton referred to such light quanta which acted like particles as *photons*. He received a Nobel Prize in 1927 as a result (sharing it with Wilson, who had invented the cloud chamber, page 29).

Just to make things even, it turns out that ordinary particles can behave as though they, too, were made up of waves. This was first pointed out in 1925 by a French physicist, Louis Victor de Broglie, who worked it out in theory.

According to his figures, an electron, for instance, should have waves associated with it that are just about as long as the waves of x-rays. This was just theory, but in 1927, the American physicist, Clinton Joseph Davisson, succeeded in detecting these electron waves. At the same time, an English physicist, George Paget Thomson (the son of J. J. Thomson, discoverer of the electron), succeeded in performing the same feat. As a result de Broglie received a Nobel Prize in 1929, while Davisson and Thomson shared one in 1937.

Since electrons have waves associated with them they can be treated like light in some ways, even though the electron waves are not electromagnetic and the light waves are.

Ordinary light can be focussed by means of lenses in such a way as to enlarge tiny objects. In this way, an instrument such as the *microscope* can be built. In the same way, electrons can be focussed by means of magnets to enlarge objects far more than any ordinary microscope using light can.

The reason for this is that any form of wave radiation can reveal only such objects as are not smaller than their own waves. If the object is smaller, the waves "step over them," so to speak. The electron waves are much shorter than ordinary light waves, and smaller objects can be seen by the former. Powerful *electron microscopes* are now very important to scientists studying very small objects.

A New Kind of Energy

There was one thing about radioactive substances such as uranium that puzzled scientists a great deal: their production of energy. Every nucleus that breaks down gives off energy. In the first place, there is the energetic gamma ray it produces. Secondly, when alpha particles or beta particles are thrown out, they move very rapidly. An alpha particle often moves at nearly one-tenth of the speed of light. That is thousands of times as fast as our fastest jet planes. Beta particles travel even faster, up to nine-tenths of the speed of light. To move even small objects at such speed takes a lot of energy.

This rapid motion means that alpha and beta particles can penetrate matter. The alpha particles can smash their way through the molecules contained in a thickness of 1 to 3 inches of air, but can be stopped by a sheet of paper. Beta particles, which are smaller but move more quickly, can smash through paper easily, and even through $1/5$ of an inch of aluminum. The heavier the atoms of a substance, the more quickly they will stop energetic particles. Lead, with very massive atoms, is a favorite substance out of which to make containers holding radioactive material. It is a *shield* against harm.

Madame Curie showed that a pound of uranium radiates away as much energy in three days as you can get by burning a millionth of an ounce of gasoline. That may not sound like much, but uranium keeps delivering that energy days without end, almost. It keeps it up year after year, century after century. By the time a billion years have passed, one pound of uranium will have delivered as much energy as you could get by burning 5,000 pounds of gasoline. And it would still be going strong, too.

Scientists had to know where all this energy came from.

It had to come from somewhere. All through the early nineteenth century experiments had made it more and more certain that energy can be changed from one form to another but cannot be destroyed altogether. Nor can it be created.

This rule is perhaps the most important "law of nature" ever discovered. The rule was first clearly stated in 1847 by the German physicist, Hermann Ludwig Ferdinand von Helmholtz, and he therefore usually gets the credit for having announced the *law of the conservation of energy*.

A similar rule seems to hold for matter. A candle may seem to disappear when it burns, but its atoms merely become a part of various gases and spread out through the air. Water seems to disappear when it evaporates, but it has merely become a vapor that spreads through the air. Iron grows heavier as it rusts. Matter isn't being created, however. Atoms of oxygen from the air are simply attaching themselves to the atoms of the iron.

In short, matter can be changed from one form to another, but it can't be destroyed or created. This is known as the *law of the conservation of matter*.

It was first worked out by the French chemist, Antoine Laurent Lavoisier, in the 1780's. Lavoisier's work in this respect is one of the important reasons why he is commonly known as the "father of modern chemistry."

Indeed, these two laws of conservation were the basis on which all of chemistry and physics was built. If either of them proved to be false, scientists would be in real trouble. Yet all that energy coming out of radioactive substances such as uranium made it look as though energy were being

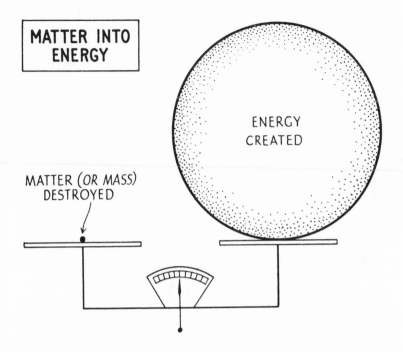

MATTER INTO ENERGY

ENERGY
CREATED

MATTER (OR MASS)
DESTROYED

created. It looked bad for the law of the conservation of energy.

The matter was straightened out by Albert Einstein, perhaps the greatest scientist of the last century. He had worked out an explanation of the photoelectric effect (see page 79) and earned a Nobel Prize for that. Some other work of his made it possible to measure the size of atoms for the first time. The French physicist, Jean Baptiste Perrin, made use of Einstein's work to calculate the size of atoms and, as a result, he was awarded a Nobel Prize in 1926.

More important still, however, was that Einstein, in 1905, devised a completely new way of looking at the universe. This is his *theory of relativity.*

As a consequence of his theory of relativity, Einstein was able to show that matter and energy are really different forms of the same thing. Matter can be "destroyed" after all, but when that happens a certain amount of energy is "created." Energy can be "destroyed," too, but when that happens a certain amount of matter is "created."

But how much is "a certain amount"? Einstein showed that when matter is converted into energy, you can calculate the amount of energy formed by multiplying the amount of matter destroyed by a figure representing the speed of light and then multiplying the result a second time by that same figure. Since the speed of light is very great, even a small amount of matter multiplies out to a tremendous amount of energy.

If a single ounce of matter were completely converted into energy, as much energy would be formed as if 60,000 tons of gasoline were burnt!

Naturally, it works the other way round. If even a large amount of energy is converted into matter, only a small amount of matter is formed. The energy obtained from burning 60,000 tons of gasoline could be turned into only one ounce of matter.

Now we have the answer to the radioactivity problem. When an atomic nucleus, such as that of uranium, breaks down, a small amount of the mass of the nucleus is changed into energy. The amount of mass lost is so small that it would never be noticed without the use of special methods. In a billion years, for one ounce of uranium, it amounts to only 1/25,000 of an ounce. However, even this mass is large enough to form gamma rays and to speed alpha and beta particles on their way at high speed.

Scientists no longer talk about matter and energy separately. There is now just the one conservation law: the *law of the conservation of matter-energy*.

Once the relation between matter and energy was understood, other things could be explained — for example, sunshine. For many years, scientists had been puzzled by the energy of the sun. There was all that light and heat that could be strongly felt even though our planet, earth, was over ninety million miles away. What was more, the sun had been delivering that energy for billions of years. All sorts of theories were made up, but none could explain why the sun hadn't run out of energy long ago.

Now came the answer. The sun is converting matter into energy. Though a small amount of matter can turn into a tremendous amount of energy, the sun's energy output is so enormous that it must use up 4,200,000 tons of matter every second to keep going. Every second the sun loses that much of its mass.

But don't let that disturb you. The sun is so massive that, even at that rate, it has enough matter to last for about thirty billion years.

ATOMIC LIFETIMES

Radioactive Slow-Down

Let's go back to a statement we made in the previous chapter: that a pound of uranium would still be going strong after a billion years. You may wonder how we can tell. After all, nobody has ever watched a piece of uranium for a billion years.

Nevertheless, scientists are quite sure of it. As a result of careful measurements with special instruments it can be calculated that in a pound of pure uranium well over a billion uranium atoms are breaking down every second. However, the total number of atoms in a pound of uranium is tremendously large. Even if the pound kept breaking down a billion atoms a second indefinitely, it would still take about thirty million years before the last uranium atom would go.

Actually it takes much longer than thirty million years because the number of atoms breaking down decreases. Let's see why this is so.

Suppose we consider something that may be familiar to most or all of you – a 1 percent sales tax. Under such a tax, whenever people buy a piece of merchandise, they must pay 1¢ for every $1 the merchandise costs. If you were to buy a television set for $200, there would be a $2 tax. If the price were reduced to $100, there would be only a $1 tax. A record-player for $50 would involve a 50¢ tax. A $10 wrist-watch band would require a 10¢ tax.

The less an object costs, you see, the smaller the tax. You can imagine that, if a store held a special kind of sale in which the price of washing machines went down $5 every day they remained unsold, the tax would go down 5¢ a day.

The way uranium atoms break down is something like that sales-tax situation. During each second, one uranium atom out of every million billion breaks down. As the atoms break down, the number of whole uranium atoms left grows smaller. As the number of whole atoms grows smaller, the number of those that break down grows smaller – just as the sales tax goes down with the price of those washing machines on sale.

Suppose the uranium broke down until there was only a half-pound of uranium left in the original lump. It would then be breaking down only half as fast as it was originally. By the time only a quarter-pound of uranium was left, it would be breaking down only a quarter as fast. As the uranium got closer and closer to the vanishing point, it would break down more and more slowly.

It is as though we were on a train rolling toward a city a hundred miles away at a speed of a hundred miles an hour. If that speed were kept up, it would take the train exactly one hour to reach the city. But suppose that as the train travelled, it lost speed constantly, so that when it was fifty miles from the city it was travelling only fifty miles an hour. At that speed it would still take an hour to cover the remaining distance.

If the train continued losing speed, the situation would never get better, either. Twenty miles from the city it would be moving at only twenty miles an hour, and it would still take an hour to get there. Five miles from the city it would be moving at only five miles an hour, and it would still take an hour to arrive. In fact, if this went on, you would suspect that the train would simply never reach the city at all, and you would be right.

Uranium breakdown proceeds in just such a way. The lump of uranium, it seems, will never completely vanish, just as the train will never reach the city.

The Long-Lived Atoms

Scientists cannot tell when the very last atom of uranium will finally go, and they don't worry about the total "lifetime" of uranium. Instead, they ask: How long will it be before exactly half of all the uranium atoms in a certain piece of uranium break down? This length of time is called the *half-life*, a term introduced by Rutherford in 1904.

The most convenient thing about the half-life is that it is the same for any piece of uranium, any piece at all. If you compare a one-ounce piece of uranium and a ten-ounce piece, you may think that the ten-ounce piece ought to take ten times as long to be reduced to half of the original quantity. After all, there are ten times as many atoms to be broken down in the ten-ounce piece. But remember the sales-tax situation. The ten-ounce piece has ten times as many atoms, true, but it is also breaking down ten times as many atoms each second. The two pieces get to the half-way point at exactly the same time.

The half-life of uranium turns out to be 4,500,000,000 (four and a half billion!) years. This time is not observed, of course, but calculated, once the number of atoms breaking down each second in any piece of uranium is known. It takes four and a half billion years for half of the uranium in any piece of uranium to break down. Then it takes another four and a half billion years for half of the half that is left to break down. Then it takes another four and a half billion years for half of the half of the half that is left to break down, and so on (like the train moving more and more slowly as it approaches the city and never quite reaching it).

When we speak of the half-life of uranium, we mean, of course, the half-life of a particular uranium isotope, *uranium-238*. Most of any piece of natural uranium is made up of that isotope. Out of every 1,000 uranium atoms, 993 are uranium-238.

What about the remaining 7 atoms out of every 1,000? These are a second isotope, *uranium-235*, sometimes called *actino-uranium*. It contains 92 protons in its nucleus, as does uranium-238, but it contains only 143 neutrons in its nucleus, three less than those in the nucleus of uranium-238.

Isotopes of the same element, we know, are very similar in their appearance and behavior. That is because the number and arrangement of the electrons in isotopes of a single element are the same. But what about things that depend entirely upon the make-up of the nucleus? The mass of an atom, for instance, depends only on the number of protons and neutrons in the nucleus, not on anything about its electrons, and we know that the mass numbers of isotopes of the same element are different. Well, the half-life of an isotope also depends on the make-up of the nucleus and not on the electrons. Two isotopes of a single atom can differ in half-life just as they differ in mass number.

If one isotope is less stable than another, it breaks down faster, and its half-life is shorter. The nucleus of uranium-235 is not as stable as that of uranium-238. In any quantity of uranium-235 there are six times as many atoms breaking down each second as in the same quantity of uranium-238. For that reason, the half-life of uranium-235 is only one-sixth as long as the half-life of uranium-238. It is only 700,000,000 (seven hundred million) years.

Now the earth is billions of years old. The most recent measurements, in fact, make it about 4,500,000,000 (four and a half billion) years since the earth's crust was formed. In all those years, there was just time for about half of the original uranium-238 to break down.

The situation is a little worse for uranium-235. It breaks down much faster; since the earth's crust first hardened, most of it has gone. Only one atom of uranium-235 out of every thirty-five still exists. Originally, two hundred and forty-five out of every thousand uranium atoms were uranium-235. They have broken down so much faster than uranium-238 that now, as we have said, only seven out of every thousand uranium atoms are uranium-235.

The radioactive element, thorium, as it is mined from the earth, consists of only one isotope, thorium-232. This isotope is even more nearly stable than uranium-238. The half-life of thorium-232 is three times as long as that of uranium-238; it is 14,000,000,000 (fourteen billion) years. In the time since the earth's crust was first formed only one out of every

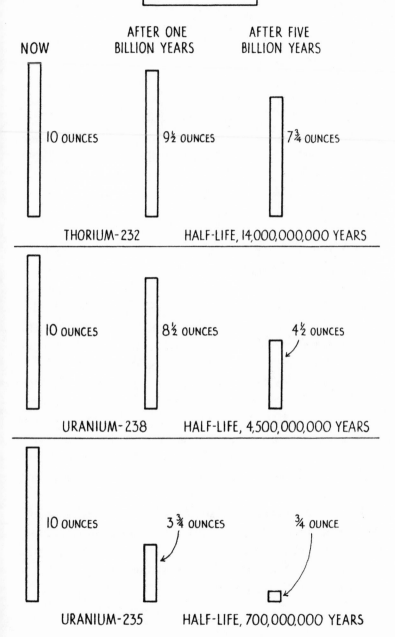

HALF-LIFE

NOW

AFTER ONE
BILLION YEARS

AFTER FIVE
BILLION YEARS

10 OUNCES

9½ OUNCES

7¾ OUNCES

THORIUM-232   HALF-LIFE, 14,000,000,000 YEARS

10 OUNCES

8½ OUNCES

4½ OUNCES

URANIUM-238   HALF-LIFE, 4,500,000,000 YEARS

10 OUNCES

3¾ OUNCES

¾ OUNCE

URANIUM-235   HALF-LIFE, 700,000,000 YEARS

six original thorium-232 atoms has had a chance to break down. Five out of six remain in existence today.

The Short-Lived Atoms

As you can see, in order for radioactive isotopes to remain in existence during all the time since the earth's crust was first formed, they must have long half-lives. Uranium-235, with a half-life of less than a billion years, is almost all gone. If a radioactive isotope had a half-life of less than half a billion years, there would be so little left of it by now that we could consider it practically gone.

Yet there are isotopes in the earth's crust today that have half-lives of much less than half a billion years.

This fact was made clear by Madame Curie and her husband, Pierre Curie. They were interested in working with minerals containing uranium compounds. In studying the radioactivity of these minerals they found that occasionally they found more radioactivity in the mineral than seemed possible. Even if the mineral were solid uranium, it ought not be as radioactive as all that.

Perhaps, thought Madame Curie, there were elements present in the mineral that were even more radioactive than uranium. If so, those elements could only be present in tiny amounts, for chemical analysis revealed no unknown elements. If the elements were present in tiny amounts and still produced so much radioactivity, they must be extremely radioactive.

The Curies decided to search for these new elements. They obtained a large quantity of a uranium-containing mineral called *pitchblende* and worked with it for months. Slowly, they discarded the non-radioactive portion and tracked down the radioactivity.

In July of 1898, they obtained a small pinch of powder containing a new element many times as radioactive as uranium. They named the element for Poland, Madame Curie's native land, and called it *polonium*. In December, 1898, they obtained a small quantity of an even more radioactive element. Because of its intense radioactivity, they named it *radium*.

It was for this that the Curies shared a Nobel Prize with Becquerel in 1903. For further work, Madame Curie received a second Nobel Prize in 1911. (Her husband had died in a traffic accident in 1906 and could not share the prize with her.)

The new elements had short half-lives. Radium (atomic number 88) has a number of isotopes, each with a different half-life.

The most long-lived isotope of radium, radium-226, was the one the Curies had detected. Even so, radium-226 has a half-life of only 1,620 years.

This isn't exactly a short half-life compared with a human lifetime. If you owned an ounce of radium and lived a hundred years, you would still have 95 percent of it left. Compared with the age of the earth, however, 1,620 years is nothing at all. There should be no radium left on earth. Yet there is.

Polonium (atomic number 84) is an even more extreme case. Its most long-lived isotope, polonium-209, has a half-life of only 100 years.

There are elements that are even shorter-lived than polonium. In 1939, a French physicist, Marguerite Perey, discovered evidence for the existence of a new element with atomic number 87. She named it *francium,* after her native France.

Its most stable isotope, francium-223, has a half-life of only 21 minutes! Yet there is reason to believe that very small quantities of francium exist in the earth.

Why is this? If elements like radium, polonium, and francium had been formed once and once only, at the time the earth was first formed, they would have been gone long ago. They would be gone no matter how much existed to begin with. If the whole earth had been made of pure francium-223 to begin with, it would all have been gone in two and a half days.

To explain the fact that these elements still exist today, we can only suppose that they are continually being formed in the earth. They are being formed today, this minute! Let us see how.

Atomic Arithmetic

When the uranium-238 nucleus breaks down, it throws off an alpha particle. What is left when the alpha particle is gone?

The alpha particle consists of two protons and two neutrons. When the uranium-238 nucleus has lost an alpha particle, it has lost two protons. Its atomic number is therefore decreased by 2 and is now 90. Atoms with number 90 are thorium. Furthermore, the uranium-238 nucleus lost two neutrons also. Its mass number is therefore reduced by 4 and is now 234.

We see, then, that when a uranium-238 nucleus throws off an alpha particle, it becomes a nucleus of thorium-234.

Don't confuse thorium-234 and thorium-232. Thorium-232 occurs naturally in the earth and has a half-life of fourteen billion years. Thorium-234, on the other hand, occurs only because it is formed from uranium-238, for it is very unstable. It has a half-life of only twenty-four days.

WEST HILLS COLLEGE LIBRARY
COALINGA, CALIFORNIA

(Quite a difference between two isotopes of the same element!)

This sort of atomic arithmetic was first worked out in 1902 by Rutherford and by an assistant of his, the English chemist, Frederick Soddy. It was this sort of calculation that showed there must be two forms of thorium at least, and Soddy therefore advanced the theory of isotopes, which Thomson and Aston were later to establish by means of the mass spectrograph (see page 65). Soddy received a Nobel Prize in 1921 as a result.

But let's carry on. When uranium-238 breaks down, it turns into something which is also unstable and which breaks down in its turn.

The thorium-234 nucleus breaks down by throwing off a beta particle. As we have seen, this means that a neutron in the thorium-234 nucleus has changed into a proton. The thorium-234 nucleus, by gaining a proton in this way, becomes 1 higher in atomic number. The atomic number is now 91, and it becomes the element *protactinium*. The mass number hasn't changed, since the only loss is an electron (the beta particle, you know), which has a mass number of almost zero. The new isotope is therefore protactinium-234.

Protactinium-234 isn't stable either. Its half-life is only a little over a minute. It breaks down by losing a beta particle. Another neutron becomes a proton, and the atomic number is up to 92. We're back at uranium, but this time the isotope is uranium-234.

Uranium-234 loses an alpha particle, and we're down to thorium-230. (You can work that out for yourself.) Thorium-230 also loses an alpha particle, the atomic number goes down from 90 to 88 (which is radium), and the mass number goes down from 230 to 226. The new nucleus is that of radium-226. That is the one that was isolated by the Curies.

As radium breaks down, it loses an alpha particle. The atomic number of what is left is now 86, and that represents the element called *radon*. Radon is a radioactive gas! It is similar to the gases helium and neon and is chemically as inert as they are. Its half-life is not quite four days.

Where does it all end? Well, after a number of additional breakdowns, the nucleus becomes the very dull and un-glamorous element *lead*. To be exact, it becomes lead-206, which is a stable isotope. There are no further breakdowns.

The entire set of changes is an example of a *radioactive series*. This particular one we have been talking about is the uranium-238 series.

Any piece of uranium ore therefore contains not only uranium-238 but also every member of the uranium-238 series, including lead-206. Eighteen different isotopes are members of the series (see diagram, pages 104 and 105), and every one is present. Each isotope is being formed just as fast as it is breaking down. (This is called *radioactive equilibrium*.) The more unstable an isotope is, the less is present, but there is always some.

This explains why small quantities of short-lived isotopes still exist in the earth.

It also explains why uranium always seems to be throw-ing off beta particles in addition to alpha particles. Actually, uranium-238 throws off alpha particles and gamma rays only, no beta particles. Some of its breakdown products throw off beta particles, however, and these breakdown products are always present.

Another result of the presence of the breakdown prod-ucts is that uranium produces even more energy than we said in the previous chapter. We allowed for only the uranium itself, but when the breakdown products break down in their turn, they also produce energy.

The alpha particles produced by uranium-238 and also by some of the other isotopes in the uranium-238 series are really helium nuclei, as we explained earlier. Eventually, each alpha particle picks up two electrons from its surroundings and becomes a helium atom. In this way, by the time one uranium-238 atom has finally broken down to one lead-206 atom, eight helium atoms have been formed. If the uranium ore is of the proper type and in the proper place, some of this helium is trapped and remains permanently with the uranium.

You might suspect that from the helium present in the uranium minerals, we could tell how long the uranium had been breaking down. Unfortunately, helium is a very light gas and some always manages to escape. It is impossible to draw conclusions from what is left, when we don't know how much has escaped.

However, the uranium also produces lead which remains in the mineral permanently (or, at least, as long as the mineral remains solid). The American chemist, Bertram Borden Boltwood, suggested in 1905 that by comparing the amount of uranium and lead present in a mineral, we could tell how long the mineral had remained solid.

One difficulty here is that some lead might be present as lead, and not as a breakdown product of uranium. Fortunately, lead is made up of four isotopes, and one of them, lead-204, is not produced in any sort of radioactivity. By measuring the quantity of lead-204 present, you can tell how much of the lead is present in its own right. The remaining lead was formed from radioactive breakdown, and from that the age of the mineral could be determined.

From such measurements (and others involving forms of

MASS NUMBERS

ELEMENT AND ITS ATOMIC NUMBER	206	207	208	209	210	211	212	213	214	215	216	217
URANIUM-92												
PROTACTINIUM-91												
THORIUM-90												
ACTINIUM-89												
RADIUM-88												
FRANCIUM-87												
RADON-86												
ASTATINE-85												
POLONIUM-84					16				12			
BISMUTH-83					15				11			
LEAD-82	18				14				10			
THALLIUM-81	17				13							

NOTE: SOMETIMES AN ISOTOPE EMITS BOTH AN ALPHA PARTICLE AND A BETA PARTICLE.
POLONIUM-218 (#8), FOR INSTANCE, EMITS AN ALPHA PARTICLE TO FORM LEAD-214
(#10) AND A BETA PARTICLE TO FORM ASTATINE-218 (#9).

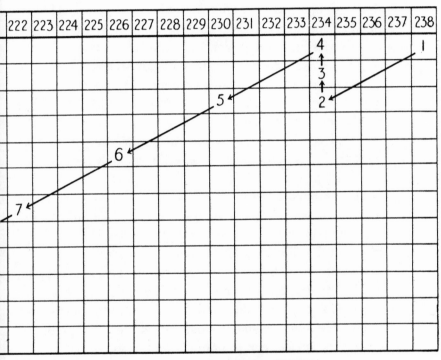

| 222 | 223 | 224 | 225 | 226 | 227 | 228 | 229 | 230 | 231 | 232 | 233 | 234 | 235 | 236 | 237 | 238 |

ARROWS LIKE THIS
INDICATE ALPHA PARTICLE
EMISSION

↑ ARROWS LIKE THIS
INDICATE BETA PARTICLE
EMISSION

radioactive breakdown) it seems quite clear that the earth's crust must have been solid for some 4,500,000,000 years. The sun and the whole solar system may have been in existence for as long as 5,000,000,000 years.

Although the uranium-238 series is the best known and most often used in age measurements, it is not the only radioactive series.

Both uranium-235 and thorium-232, the other two long-lived isotopes, are parents of radioactive series. The individual members of each series are similar to those in the other series but are not identical. A member of one of these series is not a member of either of the others. (Any piece of uranium ore also contains uranium-235 and every member of the uranium-235 series. Francium-223, which we mentioned a few pages back, is a member of this series, and that is why francium exists in nature even though its most stable isotope has a half-life of only twenty-one minutes.)

The uranium-235 series and the thorium-232 series also end up in lead, but not in the same lead isotope as the uranium-238 series. The uranium-235 series ends in lead-207, and the thorium-232 series ends in lead-208.

Except for uranium and thorium, no element with an atomic number higher than 83 has any isotope with a half-life long enough for the isotope to be existing on its own today. The other elements exist only if they are produced during the breakdown of uranium and thorium.

Radioactivity at 83 and Below

We have said that all elements with atomic numbers over 83 are radioactive. They have no stable isotopes. The next question is: Do all elements with atomic numbers of 83 or below have stable isotopes? Most of them do, certainly. But do all of them?

Well, by 1925, all the elements from 1 (hydrogen) to 83 (bismuth) had been identified in the earth's crust, with only two exceptions. Each of the elements in that list from 1 to 83 seemed to have at least one stable isotope. The missing elements were those with atomic numbers 43 and 61. There seemed to be no reason why they shouldn't have stable isotopes also. They were just very rare elements, perhaps, and had to be searched for very painstakingly. Chemists were always looking for them, therefore. Occasionally, some chemist would report having detected one or the other.

Most lists of elements in the early 1930's, for instance, included element 43 under the name "masurium" (after a district in Germany) and element 61 under the name "illinium" (after Illinois). Both often had a question mark after their names, indicating that scientists weren't certain.

By 1940 it seemed certain that the reports of earlier discoveries must have been mistaken. For some reason, neither element 43 nor element 61, as it turned out, had any stable isotopes at all. The isotope of element 43 with the longest half-life is one with mass number 99. Its half-life is 200,000 years. Element 61 is much less stable. Its most long-lived isotope (mass number 145) has a half-life of 30 years.

Neither one, naturally, could exist in the earth's crust unless it were formed from some long-lived radioactive isotope. Neither one is, so neither one exists in the crust.

You may say: Hold on! If neither element exists, how do we know anything about their isotopes, half-lives, and so on? For an answer to that, we must wait a bit longer. I will explain this point later in the book.

Let us look now at element 19, *potassium*. Potassium is one of the dozen commonest elements in the earth's crust and one of the dozen commonest in living creatures as well. One percent of the human body is potassium. (In other words, a hundred-pound boy contains a pound of potassium.) Many common chemicals contain potassium. It would certainly seem that there was nothing unusual about it.

Potassium is made up mainly of two stable isotopes. A large part of it, 93.3 percent, is potassium-39, the nuclei of which are made up of 19 protons and 20 neutrons. Most of the rest is potassium-41, made up of 19 protons and 22 neutrons. The remainder is a third isotope, potassium-40. Only one potassium atom out of 10,000 is potassium-40, yet it is this isotope which is the really interesting one. Its nucleus is made up of 19 protons and 21 neutrons. Both are odd numbers, and this generally (not quite always) means that the nucleus is unstable. Certainly, potassium-40 is unstable. It is radioactive, a fact first discovered by the English physicist, Norman Robert Campbell, in 1906.

Potassium-40 is the lightest naturally occurring radioactive isotope in the earth's crust. It occurs, of course, in the human body also. The hundred-pound boy we mentioned contains 1/600 of an ounce of potassium-40. This means that every human being and, indeed, every living creature is very slightly radioactive.

Potassium-40 has a half-life of 1,300,000,000 (one billion three hundred million) years. Since the formation of the earth's crust nine out of ten potassium-40 atoms have had a chance to break down, but that final one out of ten still exists.

Potassium-40 gives off beta particles. Each atom that

does this increases its atomic number from 19 to 20, which is the atomic number of calcium. The mass number does not change. Potassium-40 is thus converted into calcium-40, and calcium-40 is stable. Potassium-40 is therefore not the parent of a radioactive series, as the uranium and thorium isotopes are. It becomes stable in a single step.

But not all potassium-40 atoms break down by emitting a beta particle. In 1936, the Japanese physicist, Hideki Yukawa, showed that it should be possible for an atom to undergo precisely the opposite change. Instead of giving off an electron and shooting it outward as a beta particle, it can accept an electron from the outside and draw it into the nucleus.

Where would such an electron come from? From among the electrons circling the nucleus, naturally. Those electrons in the electron shell nearest the nucleus would be most easily captured. Since the innermost electron shell is called the *K-shell,* the capture of an electron by the nucleus is called *K-capture.*

The electron absorbed into the nucleus does not stay an electron, for there are no electrons in the nucleus, nor can there be any. Instead, the electron neutralizes the charge of a proton and then ceases to exist, for its own charge has been neutralized, too. The proton, in the process, has become a neutron.

As a result of K-capture, then, a nucleus has one of its protons changed into a neutron. Its mass number remains the same, but its atomic number decreases by one. (This is just the opposite of the situation that results when a beta particle is emitted.)

In 1938, the American physicist, Luis W. Alvarez, showed that the potassium-40 atom sometimes experienced K-capture. For every 100 atoms of potassium-40 that underwent breakdown, 89 emitted a beta particle, but 11 captured an electron.

Since potassium has an atomic number of 19, K-capture changes the atomic number to 18, and that belongs to the element, *argon*. Potassium-40 is converted by K-capture to argon-40.

So much potassium-40 has broken down in the history of the earth that the atmosphere possesses a great deal of argon-40. Argon belongs to the family of *noble gases,* most of which, chemists believe, were lost in the early ages of earth's history. As a result, all the noble gases are quite rare. Argon, however, is the least rare, mostly because of the existence of argon-40. The atmosphere is one percent argon-40, which means that many trillions of tons of that gas exist and can easily be obtained. And all of it arises from potassium-40.

There are other naturally occurring light isotopes (that is, those with atomic numbers of 83 or less) that are radioactive, perhaps fifteen of them. The half-lives are all longer than that of potassium-40 and often are thousands of times longer even than that of uranium-238 or thorium-232. That means they are very weakly radioactive. All of them become stable in one breakdown step.

It is rather difficult to tell when an isotope is completely stable and when it is very slightly radioactive. In 1951 certain scientists reported that bismuth-209 (the only naturally occurring isotope of *bismuth,* element number 83) is radioactive and produces alpha particles. (Till then it had been considered stable.) The scientists reported the half-life to be millions of billions of years. The half-life was so long and the radioactivity was so weak that a whole pound of bismuth-209 would throw off only about ten alpha particles per minute. Compare this with nearly five billion alpha particles per minute produced by a pound of uranium and its breakdown products.

Sometimes one wonders if perhaps no isotope is truly immortal any more than any human being is. Maybe, if our instruments were only delicate enough, we would find that all isotopes are at least very, very slightly radioactive.

Or, perhaps, all but one. There are reasons for suspecting that of all the isotopes, iron-56 is the most stable. Perhaps if the universe existed for trillions of trillions of trillions of years, all the atoms would finally be converted to iron-56.

6 ATOMIC BULLETS

The Alchemist's Dream

It has always been obvious that one substance can be changed into another. Thousands of years ago it was discovered that if a greenish rock is heated in a certain way, a reddish metal (copper) can be obtained from it. Other kinds of rock can be treated in such a way as to give us iron, lead, tin, and other metals.

Once obtained, metals can be altered again. Shining iron can turn to a dull, crumbly rust. Metals can be mixed to form new metals that are different in appearance and behavior. If copper is mixed with tin, for instance, *bronze* is formed. Bronze is yellower than copper and is also tougher and stronger than either copper or tin.

There are also changes that don't involve metals at all.

If grape juice is kept in a warm, dark place, it slowly changes into wine. Primitive man was aware of change all about him, and he could control some of the changes and make them go as he wished.

From what most people could see of the world about them, there seemed nothing wrong in supposing that any substance could be changed to any other substance. All one had to know was the right way to encourage the change. You had to heat the substance the right way or add the right chemicals.

In the late Middle Ages, some people were interested in studying the way one substance changed into another. These were the *alchemists,* and the study was known as *alchemy.*

Alchemists were particularly interested in ways of making gold. Gold represented wealth, and the only way to get gold was to dig it out of the ground. Unfortunately, the proper places in the ground were very few and quite hard to find.

It would be much simpler if you could take some substance that wasn't as rare as gold and change it into gold by proper treatment. Mercury is a metal but isn't yellow; sulfur is yellow but isn't a metal. Suppose they were mixed and pounded or heated or treated in some other way. Could one end up with something that was both a metal and yellow — in other words, with gold?

This process of changing other metals into gold was known as *transmutation,* and alchemists were always looking for means of bringing it about. Some of them claimed to have solved the problem, and some even demonstrated ways of manufacturing gold — but the demonstrations were always fakes.

The study of alchemy received such a bad name from

fakers who claimed to be able to manufacture gold that everything about the study became unpopular with serious scholars. Gradually the true science of how one substance could be changed into another came to be called chemistry instead.

By the early nineteenth century the atomic theory had been developed, and chemists were certain that it was quite impossible to manufacture gold under any circumstances. The change of one substance into another, they knew, was only the result of change in the way atoms were grouped.

Grape juice contains sugar. Sugar molecules (remember that we said a molecule is a group of closely connected atoms) are made up of 45 atoms. These include 12 carbon atoms, 22 hydrogen atoms, and 11 oxygen atoms. If the grape juice is allowed to stand, microscopic little plants, called yeast, break up these sugar molecules and change them into molecules of alcohol and carbon dioxide. The alcohol molecules are made up of 9 atoms each; these include 2 carbon atoms, 6 hydrogen atoms, and 1 oxygen atom. The carbon dioxide molecules are made up of 3 atoms each; these include 1 carbon atom and 2 oxygen atoms.

The atoms themselves aren't changed. The carbon atoms in alcohol are exactly the same as the carbon atoms in sugar. The same is true of the hydrogen atoms and the oxygen atoms. It is only the arrangement that is changed. (But that is enough to make all the difference between grape juice and wine.)

Copper can be obtained from a greenish rock because the rock is made up of molecules that contain copper atoms along with other kinds. Proper treatment separates the copper atoms but doesn't create them. When iron rusts, the rust still contains iron atoms, but also oxygen and hydrogen atoms. The iron atoms aren't destroyed.

The changes that we observe are like changes in patterns made up of colored threads. By using red, green, blue, and yellow threads, we can make any number of different patterns. But the red threads in one pattern will be just like the red threads in any other. We can't change the color of the red threads just by weaving them into a new pattern.

In the same way, no matter how you change atom combinations, you can't change one kind of atom into another.

When an alchemist tried to change mercury or lead into gold, that is exactly what he was trying to do. He was trying to change mercury atoms or lead atoms into gold atoms, and this was beyond his power.

The chemists of the nineteenth century felt sure that transmutation was beyond anybody's power. Atoms were changeless, they thought. They could be neither split nor altered. Transmutation, they decided, must remain an impractical dream of the alchemists.

And then came the discovery of radioactivity.

Getting Past the Electrons

Once radioactivity was discovered, it became obvious that atoms do change. Uranium gradually changes into various other elements, including radium, and finally settles down to lead. The same is true of thorium.

It might seem, though, that radioactivity was something special. In the first place, it mainly involved only a few elements with very high atomic numbers. Secondly, it seemed at first that radioactive breakdown could not be caused or altered by men. It couldn't be started; it couldn't be stopped; it couldn't even be slowed down or speeded up.

Ordinary chemical reactions are speeded up if the temperature is raised. Sometimes the speeding up is so great that a mixture of chemicals which is perfectly safe at ordinary temperatures will explode with great violence if heated. On radioactive breakdown, however, heat appears to have no effect.

Radium has a half-life of 1,620 years. It may be cooled to many degrees below zero or heated to red heat. The breakdown of the atoms continues undisturbed. The half-life remains 1,620 years. Radium can be put under pressure, or it can be treated in any other way. The half-life remains 1,620 years. It can be associated with other types of atoms in the form of different molecules. The association doesn't alter its manner of breakdown.

It looked, therefore, as though chemists wouldn't have to change their minds much. Instead of saying, "Atoms cannot be changed," they could say, "Atoms cannot be changed by man."

But that is wrong, too.

The trouble was that the old-fashioned ways of treating substances in the laboratory affected only the outermost electrons of the individual atoms.

To alter the way in which a radioactive substance breaks down, or to change one kind of atom into another, one must reach past the electrons. One must reach into the nucleus itself and shift its contents about. But the electrons, though light, are very efficient protectors of the nucleus. There are very few attacks that can get past them.

One way to get past the electrons, however, is to do as Rutherford did (see page 33) and bombard atoms with particles that are so small and move so quickly that they slip past the electrons (or force their way past). If many of these particles are fired at atoms, some are bound to hit the nucleus, and the electrons can't stop them. (The electrons are like a skilled prize-fighter's fists and arms, which can protect him against the fists of another man but are helpless against a bullet.)

Such tiny and speedy particles exist. The earth and everything on it, including you and me, are continually being bombarded by such atomic bullets. Let's stop to consider them for a minute.

I have already mentioned that speeding charged particles knock electrons out of the atoms with which they collide and form ions. Radioactive radiations — alpha rays, beta rays, and gamma rays — are all examples of *ionizing radiations*. They can be detected through the ions they form by such devices as cloud chambers and bubble chambers (see pages 29 and 30).

In the first years after the discovery of radioactivity such particle-detecting devices had not yet been invented. A much simpler instrument was used, one called the *electroscope*. This consisted of two pieces of gold leaf attached to a rod that was stuck through a stopper into a closed container.

If an electric charge were touched to the knob at the top of the rod, it would spread down the rod and into the two pieces of gold leaf. Both pieces would have taken up the same charge and would repel each other, spreading out like an upside-down V.

If the charge were allowed to leak away, the pieces of gold leaf would gradually approach each other again. If the electroscope is kept away from other objects and is surrounded by clean, dry air, there seems no reason why the charge should leak away. Once the pieces of gold leaf separate, they should stay separated.

However, when ions are present in the air near the leaves,

those ions can serve to neutralize some of the charge. The presence of ions causes the electroscope to discharge and the pieces of gold leaf to come together. The more ions present, the faster this happens.

Naturally, if radioactive materials are present, their radiations form ions and discharge an electroscope rapidly. Yet even when radioactive materials are absent, the electroscope discharges slowly. If the electroscope is shielded by sheets of lead that keep out radioactive radiations, it discharges very slowly, but it still discharges.

Apparently some radiation is present even when radioactive materials are not. Apparently, too, this radiation is very penetrating.

Scientists felt this radiation must be coming from the soil. To test this point, an Austrian physicist, Victor Francis Hess, took electroscopes high into the air by means of a balloon in 1911. The higher he went, he was sure, the more slowly the electroscope would discharge. Finally, it might stop discharging altogether as a couple of miles of air would stop all radiation from the soil, however penetrating.

He was wrong. The higher he went up in the air, the

THE ELECTROSCOPE

gold leaf

gold leaf

CHARGED

DISCHARGED

more rapidly the electroscope discharged. The radiation came not from the soil but from the heights of the atmosphere. For discovering this, Hess was awarded a share of a Nobel Prize in 1936.

The more these strange radiations were studied, the more it seemed they came from outside the atmosphere altogether. They came from the cosmos, or universe, itself, and from all directions. In 1925, an American physicist, Robert Andrews Millikan, suggested the radiation be called *cosmic rays* for that reason, and the suggestion was adopted.

(In 1911, the same year that Hess had discovered the cosmic rays, Millikan had performed delicate measurements which determined the exact size of the electric charge on the electron. For that he received a Nobel Prize in 1923.)

For a while, there was considerable argument as to whether cosmic rays were really rays. Millikan thought they were. He thought they were like gamma rays, only more energetic. It was found, however, that cosmic rays don't follow perfectly straight paths the way rays of ordinary light do. They bend, instead.

Compton (who invented the word, photon, see page 86) was chiefly responsible for this discovery. He travelled all over the world, measuring the amount of cosmic rays that reached earth in this place and that. He showed that the cosmic rays must bend in such a way that the polar regions get more of them than the tropical regions do.

Now the earth acts like a huge magnet. (That is why the needle on the compass always points north and south.) If the cosmic rays bend, it must be because they consist of charged particles whose path is altered by the earth's magnetism. That is the decision scientists have now come to.

Before they hit the earth's atmosphere, cosmic rays consist mostly of protons and nuclei of some of the lighter elements, moving at great speed and carrying a great deal of energy. This is called *primary radiation*. When they hit the atmosphere, they really go to work. They are so energetic that they smash any atomic nuclei they strike. They knock sub-atomic particles out of the nuclei and send them flying off at great speed. These particles, called *secondary radiation*, are almost as energetic as the cosmic rays.

Included in the secondary radiation were some kinds of particles that physicists had not yet met with. I will discuss these in a later chapter.

A New Kind of Carbon

What happens to the atoms that are struck by cosmic rays or by secondary radiation? Let's take an example.

The most common substance in the atmosphere is nitrogen. The most common isotope of nitrogen (996 atoms out of every thousand) is nitrogen-14. Its nucleus contains seven protons and seven neutrons.

Every once in a while, a flying neutron, knocked out of some atom by cosmic rays, hits one of these nitrogen nuclei. It enters the nucleus and remains there, and a proton goes flying out. (Something like this effect may be seen with marbles. A marble, skillfully shot, may hit another marble and send it flying while the first marble comes to a sudden and complete halt.)

What happens to the nitrogen-14 which has been struck by a neutron in this way? The nitrogen-14 nucleus has lost a proton, so its atomic number drops from 7 to 6. It is no longer a nitrogen atom, but is now a carbon atom. Although it has lost a proton, it has also gained a neutron, so its mass number remains the same. The new atom is *carbon-14.*

Ordinary carbon consists of carbon-12 (99 percent) and

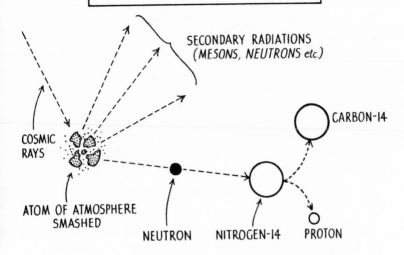

HOW CARBON-14 IS FORMED

SECONDARY RADIATIONS
(*MESONS, NEUTRONS etc.*)

CARBON-14

COSMIC RAYS

ATOM OF ATMOSPHERE
SMASHED

NEUTRON   NITROGEN-14   PROTON

carbon-13 (1 percent). Both these isotopes are stable. Carbon-14 is not stable. It is radioactive and gives off beta particles. This means that a neutron within the carbon-14 nucleus is converted into a proton. That makes it once more the stable nitrogen-14.

You may think that the net result is that nothing has occurred. However, it takes a long time for the carbon-14 to break down to nitrogen-14. Its half-life is 5,770 years, nearly four times as long as the half-life of radium. Even so, the half-life of carbon-14 would be too short to allow that isotope to exist on the earth if the isotope were not continually being formed anew by the action of cosmic rays.

You may wonder, by the way, what happens to the electrons in changes such as these. Well, they adjust themselves more or less automatically. When nitrogen-14 is converted into carbon-14, the number of protons in the nucleus decreases by one. The new carbon-14 finds itself able to hold only six electrons. There were seven electrons in the nitrogen-14 atom, so one electron is excess baggage and is turned loose. However, every time a nitrogen-14 atom is converted into a carbon-14 atom, a proton is thrown out. This proton needs to pick up an electron in order to become a hydrogen-1 atom. In this way, the electron situation is balanced. The carbon-14 atom has an electron to hand out, and the proton needs one. In any atomic breakdown or change-over, the electron situation always balances. Scientists don't worry about electrons in such situations. They concentrate only on what happens to the nucleus, and we will, too.

Once in a while, neutrons striking a nitrogen-14 nucleus produce a different change. Instead of knocking out a proton, they knock out a proton, plus two neutrons. One proton is lost, so the element left behind is carbon. Since 1 neutron comes in and 3 particles (protons and neutrons) leave, the loss in mass number is two, so that the new element is carbon-12, the common, stable isotope of carbon.

But what happens to the proton and two neutrons knocked out? They join together to form a single nucleus. Since one proton is present the atomic number is 1 and the nucleus is that of hydrogen. In fact, it is *hydrogen-3*, even more massive than hydrogen-2.

Hydrogen-3 is called *tritium* (from a Greek word meaning "third"). The interesting fact about tritium is that it is radioactive. It eliminates a beta particle and has a half-life of 12¼ years. When the beta particle is eliminated one of the neutrons in the tritium nucleus is changed to a proton. Now the nucleus consists of two protons and one neutron, and it is the stable helium-3 (sometimes called *tralphium*.)

Helium-3 exists in the atmosphere but is very rare indeed. Probably all that exists there was formed from breaking down hydrogen-3; and all the hydrogen-3 that existed was formed by cosmic ray action.

Man Takes a Hand

Atomic changes that take place as the result of cosmic-ray action don't quite fulfill the alchemist's dream, since they aren't the result of deliberate human action. Cosmic rays come from some point outside the earth's atmosphere and bombard the earth and everything upon it. There is little we can do about it. We might take an object high into the air, where cosmic rays are more frequent, so that atomic changes would be speeded up a trifle. Or we might bury an object in a deep cave, or surround it by a foot of lead, to keep out most of the cosmic rays and thus slow down atomic changes. In general, however, cosmic rays are not at the service of mankind.

But man does have other atomic bullets to serve him: the radiations from radioactive substances. True, these radiations do not carry as much energy as the cosmic rays; yet they are energetic enough to make their way past the planetary electrons and smash into the atomic nucleus.

The great advantage of such radiations is that they begin on the earth. Under neutral conditions, radioactive atoms are exploding here and there in the earth's crust, throwing out an alpha particle here and a beta particle there. That, to some extent, we can change. We can collect out of the crust the elements (uranium and thorium) responsible for the radioactivity and combine them into a single heap. We can even get out small quantities of elements like radium, which are more powerfully radioactive than uranium and thorium, and collect those into a single heap.

We can thus concentrate the radioactivity, collecting it into a powerful stream of alpha particles, beta particles, and gamma rays. Individually, the particles may be much less energetic than cosmic rays, but there are so many of them all in one place that we can accomplish much more with them.

If a piece of a radioactive substance is placed in a metal container with a small hole at one end, the metal stops most of the radiation. A thin jet of radiation, however, emerges from the hole. In that way, radiation can be aimed like bullets out of a gun.

When alpha particles are aimed, in this fashion, at a target coated with a substance called zinc sulfide, "sparks," or scintillations, result. These are quite visible and can be used to detect alpha particles and to study their behavior. (You can see scintillations for yourself if you own a wristwatch with a radium dial. The hands and numbers on the dial are coated with zinc sulfide containing the tiniest speck of radium. If you look at such a watch in the dark, you will see the scintillations struck out of the zinc sulfide by the flying alpha particles thrown off by the radium.)

During the First World War, Rutherford and some of his pupils were studying the appearance of the scintillations produced when alpha particles were passed through certain gases on the way to the zinc sulfide target. They found that something queer happened when the alpha particles passed through nitrogen. Some of the scintillations were of a kind that were produced by alpha particles passing through hydrogen. At once there was a suspicion that the alpha particles had smashed into the nitrogen nuclei and that hydrogen was produced in the process.

This turned out to be so. To check the fact, an English physicist, Patrick Maynard Stuart Blackett, took thousands of photographs of particle tracks in a cloud chamber. He collected 400,000 tracks of alpha particles and found exactly

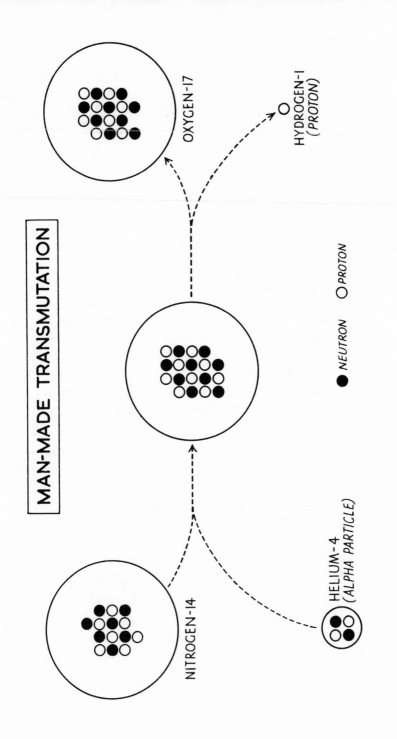

MAN-MADE TRANSMUTATION

OXYGEN-17

NITROGEN-14

HYDROGEN-1
(PROTON)

HELIUM-4
(ALPHA PARTICLE)

● NEUTRON   ○ PROTON

8 examples of collisions of these with nitrogen atoms. From these collisions he could show exactly how nitrogen nuclei disappeared and other nuclei took their place. In the course of this work, he introduced important improvements in the cloud chamber, and for that he was awarded a Nobel Prize in 1948. (A German physicist, Walther W. G. Bothe, also introduced important improvements in cloud chambers and received a Nobel Prize in 1954 as a result.)

What Rutherford figured out must have happened, and what Blackett showed actually did happen, was this: Occasionally, an alpha particle struck the nucleus of a nitrogen atom and merged with it. Shortly after the merger, the combined nitrogen-alpha particle gave off a speeding proton. It was this proton (which, after all, was a hydrogen-1 nucleus) that caused the hydrogen type of scintillation.

Now is the time for a little more arithmetic. We begin with a nitrogen-14 nucleus, which contains 7 protons and 7 neutrons. Add to that an alpha particle, which contains 2 protons and 2 neutrons. The combination of the two is a new nucleus (sometimes called a *compound nucleus*, because it is compounded of two smaller nuclei that have come together) which contains 9 protons and 9 neutrons. That combination loses a proton. What is then left behind is a nucleus containing 8 protons and 9 neutrons. This nucleus changes no further. It is oxygen-17. Oxygen-17 is one of the stable isotopes of oxygen. One oxygen atom out of every 2,500 is oxygen-17.

What has happened, then, is that by bombarding nitrogen atoms with alpha particles we have changed nitrogen into two other elements. The alpha particles are really the nuclei of helium atoms, so we can make the results of the experiment sound like a problem in addition, as follows: Nitrogen-14 plus helium-4 gives us oxygen-17 plus hydrogen-1.

This is transmutation. Elements are converted into other elements as the result of human action. The alpha particles, of course, are produced by nature, but man collects them and aims them.

And so the alchemist's dream was finally accomplished in 1919. In that year Rutherford first announced the results of his experiments, though Blackett didn't present his proof till 1925.

To be sure, the alchemists might have been disappointed in some ways. The number of atoms transmuted in this fashion is very small. Furthermore, the whole purpose of the alchemist was to make something costly out of something cheap. In this case, cheap nitrogen is being changed into equally cheap oxygen and hydrogen. Worse than that, very costly alpha particles must be used to bring about the change. Worst of all, most of the alpha particles are wasted. Only one alpha particle out of 300,000 gets around to transmuting a nitrogen atom. All the rest simply fail to hit any of the nitrogen nuclei in the proper way. They either bounce off or pass by.

Just the same, the mere fact that man can do this sort of thing at all is worth more to scientists than untold quantities of gold. And, as we shall see, it has pretty important consequences for the average man, too.

7 ATOMIC ARTILLERY

New Bullets

When an atomic nucleus is struck by a sub-atomic particle and is converted into another kind of nucleus as a result, we speak of the event as a *nuclear reaction*. The change of nitrogen-14 into oxygen-17 by the use of alpha particles is an example of a nuclear reaction. It was the first for which man was responsible.

After this first one, the men in Rutherford's laboratory proceeded to work out others. Alpha particles had smashed into the nitrogen nucleus; why not into other nuclei as well?

By 1926, about ten of the lighter elements up to potassium (atomic number 19) had been successfully transmuted by alpha particles. In every case the nucleus absorbed the alpha particle and threw out a proton.

After that things became difficult. An alpha particle contains two protons and therefore has a charge of $+2$. An atomic nucleus has a positive charge equal to the number of protons it contains. Like charges, as we saw in the first chapter, repel one another. An atomic nucleus, for that reason, repels an alpha particle.

To be sure, the alpha particle moves so fast that it smashes into some nuclei despite the repulsion. However, as the charge on a nucleus increases, the repulsion grows stronger and stronger. When the charge is $+7$, as in nitrogen, the alpha particle is still fast enough to hit the nucleus anyway. When it is $+19$, as in potassium, the alpha particle is less effective; it can barely get through. When the charge is over $+19$, the alpha particle can no longer make it. The repulsion is so strong that it cannot reach the nucleus. It is either turned aside or bounced back.

It is as though your class were members of a football team trying to score a touchdown. The opposing football team tries to prevent that. (The two teams "repel" each other.) Your team may manage to make touchdowns, anyway, despite the efforts of the other team. If the men on the other team become bigger and heavier, touchdowns become harder to make. Finally, if the opposition is big and heavy enough, touchdowns become impossible for your team. So it is with alpha particles.

Beta particles are even worse. Being electrons, they are too light to get very far (like a team of midgets playing football with Notre Dame) unless they are moving extremely fast. Furthermore, since they are repelled by the electrons in the outer reaches of the atom they usually cannot even get near the nucleus. Gamma rays are also less effective than alpha particles in bringing about nuclear reaction.

There seemed to be nothing to do but find new bullets. One possibility, first suggested by the Russian-American physicist, George Gamow, in 1928, was to use speeding protons. A proton is massive enough to smash past the electrons. What's more, its charge is only $+1$; so there is only half as much repulsion between protons and atomic nuclei as there is between alpha particles and atomic nuclei.

We could obtain protons by ionizing hydrogen — that is, by stripping the hydrogen atom of its single planetary electron. That would leave its nucleus (a single proton). The ionization could be accomplished by well-known methods. The problem was to get the protons to move quickly enough to make decent bullets.

One way to make protons (or any charged particle) move quickly is to subject them to large electric forces. For instance, if the electrons that form an electric current are concentrated in one place (or withdrawn from one place), a large charge is built up. It is a negative charge if the electrons are concentrated, a positive one if they are withdrawn. The space about such areas of great electric charge is under a *high electric potential*.

When protons are subjected to a high electric potential, they move. Their movement is toward the area of charge if the charge is negative, away from it if the charge is positive. In either case, the proton moves more and more quickly as it remains under the influence of the high electric potential. The proton is *accelerated*, or speeded up. Such protons are examples of *accelerated particles*.

You can compare the situation we are describing to man's use of water power. The falling water in natural waterfalls is made to strike a waterwheel or turbine, which turns as a result of the impact. The turning wheel can be used to run an electric generator. If the equipment is large enough,

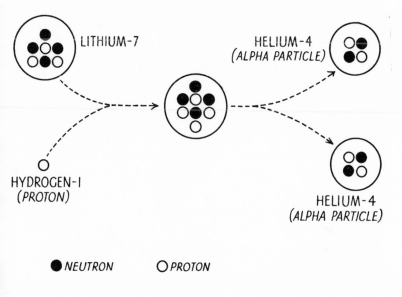

LITHIUM-7

HELIUM-4
(ALPHA PARTICLE)

HYDROGEN-I
(PROTON)

HELIUM-4
(ALPHA PARTICLE)

● NEUTRON   ○ PROTON

whole cities can be supplied with electricity from the energy of one big waterfall. The Niagara Falls are used to electrify Buffalo (New York) and its vicinity.

When no waterfall is available, it is sometimes possible for man to make one. He builds a dam across a river, and the river backs up behind it until the water forms a large artificial lake. Eventually the surface of the lake reaches the top of the dam and spills over. The effect is that of a natural waterfall. The Hoover Dam in Arizona and the Grand Coulee Dam in Washington State are examples of this.

Streams of alpha particles given off by radioactive substances are like natural waterfalls. Streams of protons and other artificially accelerated particles are like man-made waterfalls. Waterfalls, real or artificial, get their power from gravity. Accelerated particles get theirs from the electric potential.

The very first nuclear reaction brought about by accelerated particles took place in Rutherford's laboratory in 1932. Lithium-7 (its nucleus containing three protons and four neutrons) was bombarded by fast-moving protons. One proton out of every billion or so entered a lithium nucleus. The compound nucleus, which had four protons and four neutrons, instantly broke up into two alpha particles, each of which contained two protons and two neutrons.

Writing the reaction as arithmetic, we have: Lithium-7 plus hydrogen-1 gives us helium-4 plus helium-4.

This is another milestone in man's conquest of the atom. It was the first nuclear reaction in which even the bullets were man-made.

New Atom-Smashers

The successful use of protons as atomic bullets was brought about by an English physicist, John Douglas Cockcroft, and an Irish co-worker, Ernest Thomas Sinton Walton. The two shared a Nobel Prize in 1951 as a result.

To accelerate the protons to the point where they could smash into atoms with enough energy to bring about a nuclear reaction, Cockcroft and Walton built the first important *particle accelerator* in 1930. This was a device which built up electric potential step by step, multiplying it at each step. Electric potential is measured in *volts,* so the device was called a *voltage multiplier.*

The energy of a sub-atomic particle is usually measured in terms of *electron-volts.* (This is the amount of energy an electron gains if it is accelerated in an electric potential of one volt.)

In order to give you an idea of the size of the electron-volt, a molecule taking part in a chemical reaction undergoes energy changes of from 1 to 5 electron-volts. This is enough to give us the energy of such things as coal fires and dynamite explosions.

The photons of visible light have energies that vary from 1½ electron-volts for the longest waves (red light) to 3 electron-volts for the shortest waves (violet light). This is right in the range of chemical reactions, which is why so many of them produce visible light.

By contrast, nuclear reactions liberate much more energy, thousands of times as much. And to bring them about, huge amounts of energy are required. The protons accelerated by the Cockcroft-Walton voltage multiplier possessed energies of up to 400,000 electron-volts. If scientists were using atomic bullets before, they were using heavy artillery now.

Another type of device was the *electrostatic generator,* first developed by an American scientist, Robert Jemison Van de Graaff, in 1931, and often called by his name. The electrostatic generator looks like half of a dumbbell standing on end. Within it is a moving belt which continually carries electrons from the hollow bulb on top to a collection point at the bottom. As the belt moves more and more electrons from top to bottom, both the negative charge at the bottom and the positive charge at the top grow higher and higher. The higher they grow, the greater the electric potential.

The potential can be built so high that eventually a stream of electrons shoots through the air from the bottom to the top and thus neutralizes the accumulated charges. This is "man-made lightning." Natural lightning is the same thing; it results from the difference in charge between a cloud and the surface of the earth during a thunderstorm.

The electrostatic generator was much more powerful than the voltage multiplier. Indeed Van de Graaff's device was the first to produce energies of more than a million electron-volts. A million electron-volts is commonly abbreviated *Mev*. The electrostatic generator was eventually improved to the point where it produced particles with energies of 18 Mev.

The electrostatic generator was so dramatic in appearance and in deeds that it grew to be well known to the general public. It, and other devices like it, came to be called "atom-smashers."

Both the voltage multiplier and the electrostatic generator built up huge electric potentials and used these to give particles one big accelerating kick. It was also possible to give the particles a series of little kicks in such a way that each one built on the one before. (This is something like pushing someone on a swing. You don't have to give one big push. Instead, every time he swings back and begins to move forward again, you give a little push. The little pushes build up and soon the swing is moving in great arcs.)

In 1931, devices were built that could do this. The particles moved in a straight line through a series of tubes, and in each tube those particles got another little push. Such a device is called a *linear accelerator*, or a *linac* for short. As the particle moves faster and faster, each tube in the line has to be longer and longer, so that a linac can become very long indeed. Right now, a linac is being built that will be two miles long.

Still, in 1931, an American physicist, Ernest Orlando Lawrence, thought of how it might be possible to save on space. Why not make the particles move in curves instead of in straight lines?

Positively-charged particles (protons, alpha particles, and so on) are made to go round and round in a spiral under the influence of a high electric potential. As the charged particles whirl, they move faster and faster until they fly out of the container altogether. If a target is placed at the point where the flying particles emerge, nuclear reactions usually follow.

Because particles moved in circles in such a device, Lawrence named it a *cyclotron*. Even the first home-made cyclotron, only 11 inches in diameter, managed to produce particles with energies of over 80,000 electron-volts. Eventually, large cyclotrons capable of producing particles with energies of 10 Mev were built. Lawrence was honored with a Nobel Prize in 1939 as a result.

In order to produce particles of still higher energies, the cyclotron had to be modified to take care of certain queer effects at high speeds that Einstein's theory of relativity had predicted. This was first done in 1945 by two men, the American physicist, Edwin Mattison McMillan, and a Russian physicist, Vladimir I. Veksler. The modified instrument was a *synchrocyclotron*, which could produce particles with energies of 800 Mev.

An even more advanced instrument called the *proton synchrotron* can produce particles with energies in the billions of electron-volts. One such device, at the University of California, is called the *Bevatron* and another at Brookhaven National Laboratory is the *Cosmotron*. The first name comes from *Bev*, the abbreviation of "billion electron-volts." The second name implies power equal to that of cosmic rays.

By the 1960's, instruments capable of producing particles with energies of 30 Bev and more were in existence in America and in Europe. These instruments are huge objects three city blocks across, but still larger ones are being built.

The various cyclotrons are only suitable for accelerating massive particles such as protons. They won't work for electrons. For electrons, a special device called the *betatron* was developed in 1940 by the American physicist, Donald William Kerst.

Devices for producing energetic particles are not the only ones that concern modern physicists. It is necessary to detect particles also. Earlier in the book I mentioned the cloud chamber and the bubble chamber, but there are still other gadgets, too.

In 1957, the *spark chamber* was introduced. This consists of closely spaced metal plates, with alternate plates highly charged with electricity, so that an electric spark is at the point of being released. When a sub-atomic particle speeds through, sparks are released at the points where it strikes the plates.

Counters are another type of instrument used by atomic scientists. These "count" the number of sub-atomic particles that strike them. The simplest counter consists of a wire surrounded by a gas whose molecules are easily ionized by sub-atomic particles. This is enclosed in a container which has a thin spot which the particles can pierce. When a particle enters the counter, it converts the molecules of the gas into ions, allowing them to conduct electricity. A surge of electric current passes through the wire. This clicks a relay, making a little noise. The more numerous the sub-atomic particles, the steadier the chatter of the counter. To count the number of clicks, an automatic instrument called a *scaler* is used.

The first version of such a device was built in 1913 by Hans Geiger, a German physicist who had assisted Rutherford in the experiments which led to the discovery of the atomic nucleus. The device is called a *Geiger counter*, therefore.

A faster counter is one that marks off the tiny flashes of light, or scintillations, which occur when a sub-atomic particle hits certain crystals. This is called a *scintillation counter*.

New Atoms

The two nuclear reactions we have mentioned so far resulted in the formation of stable isotopes. When nitrogen-14 was bombarded with alpha particles, oxygen-17 (stable) and hydrogen-1 (stable) were formed. When lithium-7 was bombarded with protons, helium-4 (stable) was formed. The same was true of other nuclear reactions brought about in the 1920's.

In 1934, however, something new turned up. This resulted from the work of a French husband-wife team of atomic scientists, Frédéric and Irène Joliot-Curie. (The wife was a daughter of the other husband-wife team, the Curies, who had discovered radium.) The Joliot-Curies found that when they bombarded aluminum-27 (aluminum's only stable isotope) with alpha particles, protons were thrown out (as when Rutherford had bombarded nitrogen-14 in the same way), and also neutrons. In addition, however, there was a third radiation which consisted of neither neutrons nor protons. When the bombardment with alpha particles was stopped, the production of protons and neutrons also stopped, as was to be expected. (After all, if you stop using a nutcracker, you expect the nuts to stop cracking.) Yet the third radiation kept right on, even after the smashing alpha particles were removed.

Here is what happens. Aluminum-27 contains 13 protons

and 14 neutrons in its nucleus. If that nucleus absorbs an alpha particle (2 protons and 2 neutrons) and loses a proton, it gains 1 proton and 2 neutrons. The new nucleus, since it contains 14 protons and 16 neutrons, is a nucleus of silicon-30, one of the stable isotopes of the element *silicon*. But aluminum-27, when bombarded by alpha particles, also gives out neutrons. If an aluminum-27 nucleus absorbs an alpha particle and throws out a neutron, it gains two protons and one neutron. The resulting nucleus contains 15 protons and 15 neutrons. Any atom with 15 protons in its nucleus is a *phosphorus* atom. This isotope is therefore phosphorus-30.

But there is no phosphorus-30 in nature! There is only one stable isotope of phosphorus, and that is phosphorus-31, with 15 protons and 16 neutrons. Phosphorus-30 is unstable. In order to become stable, it must rearrange its nuclear contents in some way. It does this by converting one of the protons in its nucleus into a neutron. Instead of 15 protons and 15 neutrons it now has 14 protons and 16 neutrons, and it has become a stable atom of silicon-30.

This change-over from a proton to a neutron produced what I have called the "third radiation." It involves a type of particle I will discuss in the next chapter.

Phosphorus-30 has a half-life of 2½ minutes; so, after the Joliot-Curies stopped the alpha-particle bombardment, it took several minutes for the phosphorus-30 they had formed to break down. While it was doing so, the third radiation continued.

The Joliot-Curies called this *artificial radioactivity*. Phosphorus-30 was the first isotope formed by man that did not exist in nature. It was a new atom, a man-made atom. As a result the Joliot-Curies received a Nobel Prize in 1935.

Since 1934, as a result of numerous bombardments of all types of atoms with all types of natural and man-made atomic bullets, all sorts of new atoms of every element have been discovered. They are all radioactive.

The total number of radioactive isotopes created by man is well over a thousand now. You can see, therefore, that we know many more radioactive isotopes than stable isotopes. Consider, for instance, the element *cesium* (atomic number 55). Only one stable isotope of this element, cesium-133, is known, but at least twenty radioactive isotopes, ranging from cesium-123 to cesium-144, have been produced. No man-made isotopes have half-lives long enough for them to exist on the earth naturally, and none do exist naturally except a few, like carbon-14 and hydrogen-3, that are formed by cosmic rays.

Sometimes when a radioisotope is formed, the particles within the nucleus aren't arranged in the most stable way possible. When this happens, the nucleus rearranges itself into a more stable position. In doing so, it gives off a gamma ray. The original less stable arrangement is said to be an *excited state*. The final stable arrangement is the *ground state*. When a nucleus exists, thus, in two or three different arrangements, we have *nuclear isomers*. Nuclear changes from one isomer to another have a definite half-life just as do other radioactive changes. A number of stable isotopes have nuclear isomers which are unstable.

New Elements

It should be plain that if atomic scientists were going about forming radioactive isotopes of the various elements, they might some day form an isotope of an element that did not exist in nature.

In an earlier chapter, you may remember, I mentioned

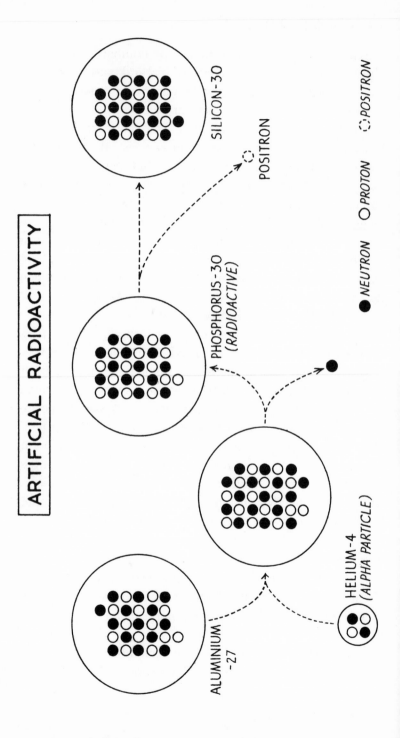

ARTIFICIAL RADIOACTIVITY

ALUMINIUM -27

HELIUM-4 (ALPHA PARTICLE)

PHOSPHORUS-30 (RADIOACTIVE)

SILICON-30

POSITRON

● NEUTRON ○ PROTON ◌ POSITRON

that elements 43 and 61 had not been definitely isolated during the 1930's. In 1937, however, an isotope of the element *molybdenum* (atomic number 42) was formed artificially by Lawrence, who used his cyclotron for the purpose. It was the radioactive isotope, molybdenum-99. The nucleus of that isotope contained 42 protons and 57 neutrons. It broke down by throwing out a beta particle, so that one neutron was converted into a proton. The resulting nucleus had 43 protons and 56 neutrons. It was the nucleus of the missing element 43. This was shown to be so by the Italian physicist, Emilio Segrè, to whom Lawrence had sent a sample of his bombarded molybdenum.

Here, then, was something really new: not just a new isotope of a well-known element, but an isotope of a new element. The new element was called *technetium,* from a Greek word meaning "artificial," because it was the first completely new element ever made by artificial methods. The particular isotope, technetium-99, has a half-life of 200,000 years, which is not long enough for it to exist on the earth naturally. (You may recall that I asked earlier how scientists could know the half-lives of isotopes of elements that did not exist. Here is the answer. Scientists make the isotope first, then measure the half-life.)

By 1942, the most long-lived isotope of element 61 had been produced and identified by a group under the leadership of the American chemist, Charles Du Bois Coryell. It has a mass number of 145, and its half-life is about thirty years. Element 61 was named *promethium,* from Prometheus, a demigod in Greek mythology who showed man the use of fire.

In addition, chemists had isolated two other elements which did exist in nature, as the result of the breakdown of uranium, but in such small quantities that the chemists knew almost nothing about them. These were francium (atomic number 87), which we mentioned before, and *astatine* (atomic number 85).

Astatine was discovered in 1940 by a group of American chemists under Segrè, the discoverer of technetium. (He had left Italy just before the war and come to the United States.) The most long-lived isotope of astatine is astatine-210, which has a half-life of only a little over 8 hours. The name "astatine" refers to this instability, since the name is derived from a Greek word meaning "unstable."

That takes care of the entire list of elements from hydrogen (atomic number 1) to uranium (atomic number 92). The next question is whether elements with atomic numbers higher than 92 can exist. The answer is yes.

Elements with atomic numbers higher than 92 are called *transuranian elements*. At the time this is written, eleven transuranian elements are known. What's more, for most of them, a number of different isotopes are known. More than fifty transuranian isotopes are known altogether.

The first transuranian element manufactured was the one with atomic number 93. It was obtained after uranium-239 was produced artificially as a result of a dramatic series of events I will describe later in the book. Uranium-239 (92 protons and 147 neutrons) breaks down by throwing out a beta particle. One neutron therefore turns into a proton, and the resulting nucleus has 93 protons and 146 neutrons.

Element 93 was first identified in 1940 by McMillan (who was later to invent the synchrocyclotron) and his associate, the American chemist, Philip Hauge Abelson.

Element 92, uranium, had been named after the planet Uranus, since the element and the planet had been discovered about the same time. Element 93 was therefore named, after the planet that was discovered after Uranus (that is, Neptune), *neptunium.*

Neptunium-239 was the first transuranian isotope isolated. It is not the most long-lived isotope of neptunium. Neptunium-237 is. That has a half-life of 2,200,000 years — not long enough for it to exist naturally on the earth.

Neptunium-237, like uranium-238, uranium-235, and thorium-232, breaks down through a long radioactive series. Unlike the other isotopes mentioned, it does not break down to a lead isotope but to bismuth-209. Since neptunium-237 no longer exists on the earth, neither do any of the isotopes into which it breaks down. On the other hand, now that neptunium-237 has been manufactured, all the breakdown products exist along with it. Here we have not only a man-made isotope and a man-made element but an entire man-made radioactive series.

Some of the neptunium isotopes break down by throwing out beta particles. Neptunium-239, for instance, breaks down in such a way. Gaining an additional proton in this manner, it becomes a new element, with atomic number 94.

McMillan, with the help of the American physicist, Glenn Theodore Seaborg, went on to identify element 94 in 1940. Seaborg continued to be very prominent in work on still higher transuranian elements. As a result, McMillan and Seaborg shared a Nobel Prize in 1951.

Element 94, named after the planet that was discovered after Neptune (that is, Pluto), is called *plutonium.* Plutonium-244 is the most long-lived plutonium isotopes, and its half-life is 70 million years.

This process of going from uranium to plutonium takes place in nature as well as in the laboratory. Occasionally, a uranium-238 atom in the soil will absorb a neutron which happens to be in the neighborhood, as a result of cosmic ray action, perhaps. The uranium atom absorbs it, becomes first uranium-239, then neptunium-239, then plutonium-239.

It has been calculated that in uranium ores there ought to be present 1 atom of plutonium for every hundred trillion atoms of uranium. In 1942, plutonium was actually detected in uranium minerals. Neptunium should also be present, but in far smaller quantities.

By 1944 two more transuranian elements had been formed. These were *americium* (atomic number 95), named for America, and *curium* (atomic number 96), named for the Curies.

In 1949, elements 97 and 98 were made. Until then, all the transuranian elements had been discovered at the University of California, located in Berkeley, and this fact was used in finding names for the new elements. Element 97 was called *berkelium* and element 98 *californium*.

In 1954, the manufacture of elements 99 and 100 was announced, and in 1955 the manufacture of element 101. Their names are einsteinium, fermium and mendelevium, after three famous atomic scientists.

In the name, mendelevium, you can recognize the Russian, Mendeléev, who had first worked out the periodic table of the elements a century before. Einsteinium is named in honor of Einstein, who had died just a few months before the element received its name. Fermium honors the Italian-American physicist, Enrico Fermi, who had also just died and whose work will be described later in the book.

In 1957, the synthesis of element 102 was announced. It was named *nobelium* after the Nobel Institute in Stockholm, where it was first formed.

Unfortunately, the work done at the Nobel Institute could

not be repeated. Other scientists formed element 102, but by other methods, not by the one described at the Institute. As a result the name, nobelium, is not official. However, no other name has been proposed.

In 1961, element 103 was formed. It was named *lawrencium* in honor of Lawrence, the inventor of the cyclotron. In 1964, element 104 was formed, and in 1967, element 105. These are named *rutherfordium* and *hahnium* after Rutherford, who discovered the atomic nucleus, and Otto Hahn, whom I will mention later in this book.

All the transuranian elements are radioactive. What's more, they get harder and harder to make, and their half-lives grow shorter, as the atomic number grows larger. The longest-lived known isotope of element 99 (einsteinium-254) has a half-life of a year and a half. The isotope of element 100 (fermium-253) that is known to have the longest life, has a half-life of four and a half days. The stablest known isotope of element 101 (mendelevium-256) has a half-life that may be as short as an hour and a half.

The isotopes of elements with atomic numbers over 101 have half-lives only a few minutes or even a few seconds long. Some physicists think, though, that there is reason to suppose that elements with atomic numbers between 110 and 114 may have fairly long-lived isotopes. Perhaps these will be made some day.

8 ATOMIC NEWCOMERS

Opposites

So far in this book, I have discussed only three kinds of sub-atomic particles: electrons, protons, and neutrons. It seems, perhaps, that these three are quite enough; that they explain the structure of atoms well, and how nuclear reactions proceed.

Actually, they are not enough, and on two different occasions I have hinted that other particles exist. I mentioned the secondary radiation produced by cosmic rays striking the atmosphere. I also mentioned a new kind of radiation detected by the Joliot-Curies when they discovered artificial radioactivity. In both cases, new kinds of particles I have not mentioned are involved. It is time now that I talked about them.

One hint at the existence of additional particles came in

1930. In that year, an English physicist, Paul Adrien Maurice Dirac, was working out mathematical equations that would describe the behavior of the electrons. It seemed to him that if the equations he obtained were correct, there ought to be two kinds of electrons. One was the ordinary electron with its negative charge of −1. Another, however, was the exact opposite of the electron; a particle with just the mass of the electron, but with an opposite charge of +1.

Could there really be such a "positive electron"? In 1932, the American physicist, Carl David Anderson, was studying cosmic rays. Cosmic rays entered his cloud chamber, which contained a lead bar, and smashed into the lead atoms. Other particles were produced, and among them Anderson noticed a trail of droplets that looked exactly as though it were made by an electron. The only trouble was that it curved the wrong way; it was positively-charged and not negatively-charged.

Anderson had discovered the positive electron that Dirac had suggested ought to exist. He named it the *positron.* Because it is the opposite of an ordinary particle, the positron is considered an example of an *anti-particle.* It is sometimes called an *anti-electron,* in fact, because it is the opposite of an electron.

Because of this, Dirac was awarded a Nobel Prize in 1933, and Anderson was awarded one in 1936.

Positrons were quickly discovered elsewhere than among the products of cosmic ray bombardment. Earlier I mentioned that the Joliot-Curies had produced the first example of artificial radioactivity in 1934. They had formed atoms of phosphorus-30, with nuclei made up of 15 protons and 15 neutrons. Phosphorus-30 is unstable and changes a proton to a neutron to become the stable silicon-30 with 14 protons and 16 neutrons.

But what happens when a proton changes to a neutron

within a nucleus? I have already explained that when a neutron changes to a proton within a nucleus, a beta particle (an ordinary electron) is emitted. The proton-to-neutron change is just the opposite, and therefore you would expect an opposite particle to be emitted. Instead of shooting out an ordinary electron, a phosphorus-30 nucleus shoots out a positively-charged electron, a positron. Since then, numerous radioactive isotopes have been formed which break down by emitting positrons.

Actually, the positron is not the only anti-particle that ought to exist. Dirac's mathematics made it seem that almost any particle ought to have an opposite. The proton ought to have an opposite, for instance, an *anti-proton*.

The trouble was that the more massive a particle the more energy was required to bring it into existence. An anti-particle would be as massive as a proton and 1,836 times as massive as either an electron or a positron. It would take 1,836 times as much energy to make an anti-proton as a positron.

Almost any cosmic ray particle had energy enough to produce a positron. Only very energetic, and very rare, cosmic ray particles could form an anti-proton.

By the time scientists had developed particle accelerators that could produce billions of electron-volts, enough energy was at hand. In 1956 Segrè (the discoverer of technetium and astatine) and a young assistant, the American physicist, Owen Chamberlain, made use of the Bevatron to produce the anti-proton and show it actually existed. As a result, Segrè and Chamberlain received a Nobel Prize in 1959.

The anti-proton was as massive as the proton but carried a negative charge (−1) instead of a positive charge.

Oddly enough, it is also possible to form an *anti-neutron*. The anti-neutron has the same mass as the neutron and the same lack of charge. Both neutron and anti-neutron have a charge of 0; are uncharged, that is. How, then, is one the opposite of the other?

This comes about because most sub-atomic particles act as though they are spinning about their axis. When a charged particle spins in this way, it sets up a magnetic field which has a definite direction. The north magnetic pole points in one direction, the south magnetic pole in the other. In an anti-particle, the magnetic field is reversed as compared to the particle. The north magnetic pole of an electron or proton points upward, for instance, while the north magnetic pole of a positron or an anti-proton points downward.

The neutron, even though it is uncharged, sets up a magnetic field when it spins. The anti-neutron has the same mass and lack of charge the neutron has, but its magnetic field is pointed in the opposite direction.

Anti-Matter

We now have two sets of particles. The first set is made up of the old familiar particles: the proton, neutron, and electron. The second is made up of their opposites: the anti-proton, anti-neutron, and positron.

Ordinary atoms are made up of the first set, but we can imagine atoms made up of the second set. We can imagine atomic nuclei made up of anti-protons and anti-neutrons. Such nuclei would carry a negative charge instead of a positive one. To balance the negative charge, such nuclei would be surrounded by positrons.

We would have *anti-atoms*, then, made up of anti-particles only. A collection of anti-atoms would make up *anti-matter*.

Can such anti-matter actually exist? In 1965, physicists

managed to put together an anti-proton and an anti-neutron to form a nucleus of *anti-deuterium*. Later, Soviet scientists prepared *anti-helium-3* (two anti-protons and an anti-neutron). That is as far as anyone has gone so far. Doing even that much was difficult.

The trouble is that anti-particles cannot last long. Once a positron is formed, for instance, it finds itself surrounded by uncounted numbers of electrons. When any particle meets its anti-particle, they cancel each other, so to speak. They undergo *mutual annihilation*, where a particle and anti-particle existed before; no matter at all exists afterward.

After a positron is formed, it takes only a millionth of a second or so for it to meet an electron, collide, and be annihilated. (Sometimes the positron and electron circle each other to form a combination called *positronium*, first detected in 1952, before actual collision and annihilation.) The same is true for anti-protons and anti-neutrons which quickly collide with protons and neutrons, once they are formed, and disappear.

This is the reason it is so difficult to work with anti-matter.

Of course, the particles don't disappear altogether after colliding. They cease to exist as matter, but they appear as energy. Einstein's theory of relativity predicts how much energy ought to be formed when the mass of an electron and positron disappear, and this amount of energy, exactly, is formed. It is also possible to form out of energy an electron-positron pair. (This is called *pair-formation*.) Again, the exact amount of energy is used up that is called for by Einstein's theory.

When a proton and anti-proton annihilate each other, 1,836 times as much mass is involved as in the case of an electron or a positron, and 1,836 times as much energy is formed. Such measurements of disappearing mass and appearing energy form some of the best evidence that Einstein's theory is accurate and useful.

Physicists are rather puzzled at the fact that the universe seems to be made up entirely of particles; and that anti-particles are so rare. There are reasons to think that particles and anti-particles ought to exist in the universe in equal quantities.

Some scientists think that perhaps there are two universes. There is our own, made up almost entirely of ordinary particles, and also an "anti-universe" somewhere, made up almost entirely of anti-particles.

Others think that our universe is made up of both matter and anti-matter, each existing in huge masses of stars called *galaxies*. Our own galaxy (the Milky Way) seems to be made up of matter, but other galaxies may be made up of anti-matter.

Astronomers are watching for special cases of collisions between matter-galaxies and anti-matter-galaxies, for these ought to produce vast floods of energies. Some galaxies do seem to produce such vast floods, and there is just a possibility that we are watching matter and anti-matter in collision on a large scale.

The Tiniest

One of the puzzling items in the study of atomic structure arose in connection with beta particles. In the first decade after the discovery of radioactivity, it turned out that beta particles were given off by particular atoms with different amounts of energy.

It seemed to physicists that a particular atom ought to give off beta particles with only a fixed amount of energy and no other. Instead, the amount of energy was always less than that which scientific theories demanded; and could be less by any amount. Some beta particles were emitted with hardly any energy.

This seemed to violate the law of conservation of mass-energy, and this was very distressing. That law was upheld in so many different ways and proved to be so useful that scientists hated to have it broken. They therefore thought up ways in which the law could be saved.

In 1931, Wolfgang Pauli (who had helped work out the system of electron-shells, see page 48) came up with a suggestion. He proposed that whenever a beta particle was produced, a second particle was also produced. The energy that was missing in the beta particle was carried off by this other particle.

Why could not this other particle be detected then? Pauli showed that if this other particle existed, it would have to be electrically uncharged, so that it could not be detected in a cloud chamber. It would also have to be very light, much lighter even than an electron. It might even have no mass at all.

This particle came to be called a *neutrino* (an Italian word meaning "little neutral one"). Naturally, a particle without charge and without mass would be very difficult to detect. The Italian physicist, Enrico Fermi, worked out the manner in which such particles ought to behave, and it turned out that they could travel through matter almost as though matter were empty space. Neutrinos could penetrate a sheet of solid lead as thick as from here to the nearest stars.

But if a neutrino had no mass and no charge, and if there were no way of detecting it (for it could only be detected if it could be stopped, and light-years of matter wouldn't stop it), how could it exist? It was a "nothing-particle." Well, not quite. The neutrino had a spin and a magnetic field. It carried energy, and every once in a long while, one of them was indeed stopped by matter.

Many physicists were skeptical that neutrinos really existed, but other physicists insisted on their existence. Not only did they offer a way out to save the law of conservation of mass-energy, but they also saved several other important natural laws which would be broken if neutrinos didn't exist.

Actually, there are two kinds of these tiniest of all particles: ordinary neutrinos and *anti-neutrinos*. When an atom emits a beta particle (a speeding electron), it is an anti-neutrino that accompanies it. When an atom emits a positron, it is a neutrino that accompanies it.

Two American physicists, Clyde L. Cowan, Jr. and Frederick Reines, were determined to track down this pair of particles. They made use of a nuclear reactor (something I will describe in detail later in the book) which they reasoned ought to be a source of uncounted trillions of anti-neutrinos.

This invisible beam of neutrinos was allowed to collide with large tanks of water, containing certain chemicals. Every once in a while one anti-neutrino, out of all those trillions, ought to collide with a proton and undergo a nuclear reaction. The water targets with their chemicals were so arranged that if such a reaction took place, gamma rays would be produced in one particular pattern and no other. In 1956, the necessary pattern of gamma rays was observed, and the anti-neutrino was actually detected, a quarter-century after Pauli had suggested its existence.

Of course, if anti-neutrinos exist, neutrinos do. In fact, the nuclear reactions that keep the sun shining are of a type that produce neutrinos.

This means that the earth and everything on it is sub-jected to a vast constant flood of neutrinos from the sun. About 60 billion neutrinos are constantly passing through each square inch of our skins every second, day or night. (Even at night, the neutrinos which strike the other side of the earth, the sunlit side, merely go through the earth and then through us — travelling always at the speed of light.) Such neutrinos do no harm to us. They merely pass through us as though we were not there.

Scientists have set up devices now that will detect neutrinos pouring down from the sun and from other stars. Such neutrinos may give us information about the center of the sun that ordinary light cannot give us. Also there are theories that just before stars explode they emit particularly large quantities of neutrinos. By spotting rich sources of neutrinos in the sky, we may be prepared to watch for star-explosions. In fact, a whole new science, *neutrino astronomy,* is developing.

In Between

Another puzzle in atomic structure involved the nucleus. By 1932, it was clear that the nucleus was made up of protons and neutrons. In that case, how did it hold together? Protons all carried a positive charge and therefore repelled each other. In the atomic nucleus, the protons were jammed very close together and since the repulsion increased the closer they were, the nucleus ought to experience very large repulsions. In fact, the nucleus ought to explode with unimaginable force; but it didn't.

Of course, the presence of neutrons seemed to hold the protons together, but how?

Attempts to work out the theory of atomic structure had to make use of Planck's quantum theory (see page 78). Bohr had been the first to do this, and his work made it possible eventually to explain the existence of electron shells.

An Austrian physicist, Erwin Schrödinger, worked out the mathematics of this *quantum mechanics* in detail in 1927, and his work was improved upon by the German-British physicist, Max Born. This was important enough to result in the award of a Nobel Prize to Schrödinger in 1933 and to Born in 1954.

Quantum mechanics was used by a German-American physicist, Marie Goeppert-Mayer, for instance, to try to work out the arrangement of protons and neutrons within the nucleus. She and a German physicist, J. Hans Daniel Jensen, worked out a system of nuclear shells in 1948, somewhat like the system of electron shells, but more complicated. This won them Nobel Prizes in 1963.

But how could quantum mechanics be used to explain how the nucleus was held together in the first place?

The answer began with the work of Heisenberg, who had been the first to suggest the proton-neutron make-up of the nucleus. He had worked out a system of quantum mechanics of his own in 1926, and in 1927 he showed that it was impossible to make certain measurements in an absolutely exact fashion. There would always be a bit of uncertainty in any measurements, and this uncertainty would be particularly noticeable where particularly small objects such as sub-atomic particles were involved. This *uncertainty principle* won for Heisenberg a Nobel Prize in 1932.

A Japanese physicist, Hideki Yukawa, decided to make use of the uncertainty principle in his study of the nucleus. He decided that there must be an attractive force between protons and neutrons that was even larger than the repulsive force that drove protons apart because of their electric charge.

This *nuclear force* was very short-ranged. It could only

exhibit itself inside the tiny atomic nucleus. Outside the nucleus it could not be detected.

Yukawa tried to work out conditions that would allow the nuclear force to be very strong inside the nucleus and very weak outside. From the uncertainty principle, he decided that there must be a new kind of particle produced within the nucleus. This new particle bounced back and forth between protons and neutrons but was so short-lived that it never lasted long enough to get outside the nucleus. Furthermore, Yukawa concluded that the mass of this new particle would have to be in between the masses of the familiar particles. It had to be more massive than an electron and less than a proton or a neutron. To be exact, it ought to be about 250 times as massive as an electron.

Yukawa announced his conclusions in 1935, and the very next year, Anderson (the discoverer of the positron), discovered such a particle among the secondary radiation produced by cosmic rays. Indeed, such a particle made up the major portion of the secondary radiation.

Anderson called the new particle a *mesotron* (from a Greek word meaning "intermediate") because it was intermediate in mass between electrons and protons. This was quickly shortened to *meson*.

Unfortunately, Anderson's meson proved not to be the particle Yukawa had predicted. It was only about 200 times as massive as an electron, and it didn't behave as it ought to. If it were Yukawa's particle it ought to be absorbed very quickly by matter, but Anderson's meson penetrated considerable thicknesses of matter without being absorbed.

In 1947, however, an English physicist, Cecil Frank Powell, discovered another kind of particle in the secondary radiation. This new particle was a bit more massive than Anderson's particle (it was about 270 times as massive as an electron), and it had all the properties Yukawa had predicted.

Since Powell's particle was formed by the primary radiation, it was given the initial "p" for "primary." However, the Greek form of "p" was used so that Powell's particle came to be called the *pi-meson*. This is often abbreviated to *pion*.

Anderson's particle was given the initial "m" for "meson." Using the Greek letter, it became *mu-meson* or *muon*.

There are two muons, a negatively-charged one and a positively-charged one. The positive muon is an anti-particle.

There are three pions: a positively-charged one, a negatively-charged one, and an uncharged one. The negative pion is an anti-particle. The uncharged pion is both particle and anti-particle at the same time.

The pion is definitely the particle Yukawa predicted. Its existence explains the nature of the nuclear force that holds the atomic nucleus together despite the repulsion of the electric charges for each other. As a result Yukawa received a Nobel Prize in 1949, and Powell received one in 1950.

Nowadays, the sub-atomic particles are divided into three groups. There are the *leptons, mesons,* and *baryons.* The leptons include the light particles, such as electrons, positrons, and neutrinos. The mesons include the particles of intermediate mass, such as the pions. The baryons include the massive particles, such as the proton, neutron, anti-proton, and anti-neutron.

Remaining Puzzles

Physicists have by no means solved all the puzzles of atomic structure. New puzzles arise as old ones are solved, it seems.

It turns out, for instance, that there are two nuclear forces. One of them, the *strong interaction,* is the one that holds the atomic nucleus together. The other, the *weak interaction,* is much weaker, and it governs the manner in which many sub-atomic particles break down.

It turns out that certain laws of nature which scientists felt were true under all conditions may be true for one type of force and not for another.

For instance, one such law is called the *law of conservation of parity.* This law makes it seem as though the universe doesn't distinguish between left and right, or between an object and its mirror image. In 1956, two Chinese-American physicists, Tsung-Dao Lee and Chen Ning Yang, maintained that this law did not hold for weak interactions and described experiments that would prove it. The experiments were tried, and Lee and Yang turned out to be right. They were awarded a Nobel Prize in 1957.

Scientists are still trying to work out the details of how weak interactions take place.

Then, too, there is the question of the muon. It was the pion that turned out to be Yukawa's particle and that left the muon nowhere. What did it do? What part did it play in the atom?

Nobody has ever found out.

What scientists have found out, though, is that the muon has a set of properties that are almost identical with those of the electron. The only important difference is that the muon is 200 times as massive as the electron. The muon, therefore, is a "heavy electron," and not much more. A muon can even take the place of an electron in an atom and circle the nucleus in its place to form a *mesonic atom.* It can circle a positron in place of an electron to form a short-lived combination called *muonium.*

Why should a muon be 200 times as massive as an electron? Why shouldn't all that mass make it more different than it is? Nobody knows.

When muons are formed, neutrinos are formed along with them, just as they are formed along with electrons. In 1962, physicists discovered that the neutrinos formed along with muons do not bring about exactly the same kind of events when they are (only once in a while) absorbed by matter, as do the neutrinos formed along with electrons.

There exists in other words a *muon-neutrino* that is different from an *electron-neutrino;* and a *muon-anti-neutrino* that is different from an *electron-anti-neutrino.* Both types of neutrinos have neither mass nor charge, and both have the same kind of spin. How, then, are they different? Nobody knows.

Since 1950, and especially since 1960, numerous mesons and baryons have been discovered. By the middle 1960's there were over a hundred different sub-atomic particles known. Why are there so many? What do they all do? Nobody knows.

Physicists are trying to group the sub-atomic particles into families, in the same way that Mendeléev once grouped elements into families in his periodic table. One interesting theory of this type was proposed by the American physicist, Murray Gell-Mann, in 1961 (and, at the same time, by the Israeli physicist, Yuval Ne'eman). According to Gell-Mann's theory, a particular particle which he called the *omega-minus particle* ought to exist. He described its properties in detail, and in 1964 such a particle was discovered. It had the properties Gell-Mann had described.

Other even more elaborate theories than Gell-Mann's were advanced in 1965. Will any of them explain the existence of all the known sub-atomic particles? As yet no one knows.

Are the sub-atomic particles all different? Or are they collections of still simpler and smaller sub-subatomic particles?

In the 1950's, the American physicist, Robert Hofstadter, bombarded nuclei with high-energy electrons. It was possible for him to see inside protons and neutrons as a result, so to speak. He decided that protons and neutrons were actually built up of mesons. (The neutron might be uncharged altogether, but it contained both positively- and negatively-charged particles. From the way these charges were distributed it was possible for the neutron to set up a magnetic charge when it spun. It is this which accounts for the existence of an anti-neutron. See page 148.)

Hofstadter received a Nobel Prize in 1961 as a result, but the inner structure of the sub-atomic particles is far from settled.

ATOMIC ENERGY

Not Exactly 1

Although there are many puzzles remaining in connection with atomic structure, much is known. That which is known has already proven of great use (and of great danger, too) to mankind.

Part of the usefulness and the danger arises from the question of the mass of protons and neutrons. All through this book I have been calling the mass number of the proton and of the neutron 1. Actually, the mass numbers are exactly 1 only in certain special cases.

For instance, the carbon-12 nucleus has a mass number which is purposely set at exactly 12. Carbon-12 is the standard for measurement of mass. All other atomic nuclei and sub-atomic particles are compared with it as far as mass is concerned. (This is similar to the way in which we compare lengths of objects with the length of a ruler. A ruler is a standard for the measurement of length.)

A carbon-12 nucleus contains 6 protons and 6 neutrons, or 12 sub-atomic particles altogether. The average mass number of the protons and neutrons in the carbon-12 nucleus is, therefore, exactly 1.

In other nuclei, however, protons and neutrons have mass numbers that are very slightly different from 1. A proton all by itself has mass number 1.0078. A neutron all by itself is just a little more massive; its mass number is 1.0087. If a proton and a neutron combine to form the nucleus of hydrogen-2 (deuterium), the average mass number of the two particles is not 1.00825, but 1.00705. In carbon-12, with 6 protons and 6 neutrons in the nucleus, the average mass number of those particles is exactly 1, as we have said. In oxygen-16, with 8 protons and 8 neutrons in the nucleus, the average mass is 0.9997.

In other words, the mass of individual protons and neutrons gets smaller as more and more of them are packed together into larger and larger nuclei. The mass continues to get smaller in elements more complicated than oxygen. In sulfur-32, for instance, the average particle in the nucleus has mass number 0.9991.

The mass number is smallest for the particles in middle-sized atoms such as iron and copper. In iron-56 the mass number of the average particle is 0.9988.

From that point on, as atoms continue to grow more and more complicated, the mass number of the particles in the nucleus starts growing larger once more in a very slow way and continues to do so to the end of the list of elements. By the time the heaviest stable isotopes are reached, the average mass of the protons and neutrons in the nucleus is just about 1, and in the case of elements such as thorium and uranium it is slightly over 1.

You may see how this works more clearly if you look at Mass-Energy Diagram I, below. See how, beginning with a high mass number for hydrogen, the line drops rapidly until it is at a low point in the region of iron, then rises more slowly to the very end.

The change in mass is exaggerated in Mass-Energy Diagram I. The most massive a proton or neutron can be, under ordinary conditions, is 1.0087. The least massive it can be is 0.9988. If we had 120-pound weights that varied in the same proportion, the variation would be a single ounce in either direction. We would probably never notice it. Or, if we did, we might shrug our shoulders and say, "What difference does an ounce make in a hundred and twenty pounds?"

In atomic affairs, however, it makes a great deal of difference.

The Disappearing Mass

What happens to the part of the mass of protons and neutrons that disappears when they are packed into middle-sized atomic nuclei?

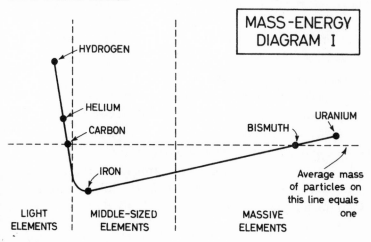

MASS-ENERGY DIAGRAM I

HYDROGEN

HELIUM

CARBON

URANIUM

BISMUTH

IRON

Average mass of particles on this line equals one

LIGHT ELEMENTS

MIDDLE-SIZED ELEMENTS

MASSIVE ELEMENTS

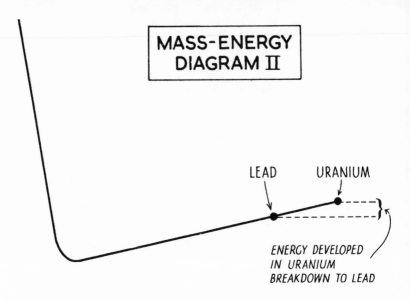

LEAD   URANIUM

*ENERGY DEVELOPED
IN URANIUM
BREAKDOWN TO LEAD*

There can be only one thing happening to any mass that seems to disappear. It is converted into energy. When one nucleus is changed into another nucleus containing particles of lower mass, energy is developed. The energy may take the form of a gamma ray, or of speeding particles, or of simple heat.

Whenever a very light atom or a very massive one is converted into some atom nearer the middle of the list, energy is developed. Uranium, for instance, gives off energy as it gradually changes to lead during its radioactive breakdown. If you look at Mass-Energy Diagram II, you will see why. The particles in the lead nucleus are less massive than those in the uranium nucleus. The change from uranium to lead is downhill on the diagram, and in this way the energy released in radioactivity is produced.

In fact, in all radioactive changes, the particles produced weigh slightly less than the original particles. For instance, in the conversion of carbon-14 to nitrogen-14, the carbon-14 begins with a mass number of 14.0033, while the nitrogen-14 produced has one of only 14.0031. In the same way, a neutron with a mass number of 1.0087 changes to a proton with a mass number of 1.0078.

The energy produced by changes in atomic nuclei is called *atomic energy*. A better name is *nuclear energy*, but the former is more common.

From the very discovery of radioactivity onward, people have wondered if there weren't some way in which this atomic energy might be put to use. There is so much of it, after all! The atomic energy that could be released in converting a pound of coal into a pound of iron is millions of times greater than the amount of energy released when coal is simply burned.

For a long time, however, there seemed to be no practical way in which atomic energy could be harnessed by man. To be sure, one type of atom could be changed into another. The trouble was that to do so you had to bombard the atoms with sub-atomic particles. For every sub-atomic particle that found its mark and changed an atom, a million or more missed. Not only was the number of atoms changed very small, but the amount of energy scientists had to put into firing all those sub-atomic particles was very great. Atomic energy seemed to be a losing proposition. Much more energy was put in than could be got out.

It was like buying dimes for a hundred dollars apiece. The dimes are money, yes. Still, the more of them you bought in that way, the poorer you would be.

One way in which it seemed that nuclear reactions could be made to waste less energy was to cut down the number of misses. To do that, a brand-new type of bombardment was employed.

The Useful Neutron

One of the reasons for the large percentage of misses that take place when sub-atomic particles are used to bombard atoms is the repulsion between the nucleus and the particle. We've talked about that already. Even when particles are accelerated to huge speeds, only a small fraction succeed in smashing through despite the repulsion. Most of them always glance off.

Now suppose we use a particle for which neither the atomic nucleus nor the surrounding electrons have any repulsion. Such a particle is the neutron, which has no charge.

When a neutron strikes an atom, it passes right through the electron shells and right into the nucleus (if it is aimed right). Negative charges don't repel it, nor do positive charges. The neutron need not even be travelling quickly. Even if it glances off a number of atoms and is slowed up in the process (just as a running man would slow up if he entered a crowd and had to jostle his way through), it can still enter an atomic nucleus if it hits one squarely. Such slow neutrons are called *thermal neutrons*.

If we go back to our comparison to a football team, we may look on the neutron as an invisible player carrying an invisible ball. An ordinary player is stopped by the opposing team when he is trying to score a touchdown. He succeeds in scoring one only rarely. Our invisible player can score a touchdown every time because the opposing team is not trying to stop him. They can't even see him. Our invisible player can even score a touchdown by simply walking across the line with his invisible ball.

In fact, there are theories that it was by neutron capture that the universe, as we know it, was originally formed. Some scientists, such as George Gamow, suggest that the universe began as an exploding mass of neutrons. Some of the neutrons broke down to protons and electrons, then built up more complicated nuclei by absorbing additional neutrons. Gamow thinks that all today's elements may have been formed within half an hour of the original explosion.

Neutrons enter some nuclei more easily than others. It depends on the arrangement of particles in a particular nucleus. A nucleus that is easily entered by a neutron is said to have a large *cross-section*. Scientists have now worked out the size of the cross-section of most nuclei for the various types of bombarding particles, particularly neutrons.

But where does one get a stream of neutrons? One can always get alpha particles and beta particles from uranium or thorium and their breakdown products. One can get speeding protons by simply ionizing ordinary hydrogen and accelerating the nuclei. There is no such simple source of neutrons.

A source had to be made, and nuclear reactions were used for the purpose. We have already mentioned the Joliot-Curies, who bombarded aluminum-27 with alpha particles and got three types of radiation. One of these types consisted of a stream of speeding neutrons. A number of other nuclear reactions also liberate a stream of neutrons in this way.

The neutron streams resulting from various nuclear reactions are quite hard to detect. Neutrons don't ionize atoms in the atmosphere the way charged particles do. In the absence of ions, the usual instruments for detecting sub-atomic particles don't work. It is necessary, therefore, to place one of certain substances in the way of what is suspected of being a stream of neutrons. The neutrons, if they are there, strike the substance, causing nuclear reactions which may liberate charged particles. These new particles then form ions and can be detected.

The substance most commonly used in this way is *boron-10*. When boron-10 is struck by neutrons, alpha particles are liberated. That is a dead give-away that neutrons are shooting about.

Another difficulty is that neutrons cannot be accelerated.

A high electric potential has no effect on neutrons since neutrons possess no electrical charge. For a while it seemed that scientists would just have to take neutron streams at whatever energies they happened to have.

One way out of this spot was to use the nucleus of hydrogen-2 (deuterium). This nucleus consists of a proton and a neutron in close association. It has a single positive charge and can be accelerated just as a proton can. The hydrogen-2 nucleus is called a *deuteron* or, very occasionally, a *deuton*.

Now suppose a deuteron is approaching an atomic nucleus. The results were worked out by the American physicist, Robert Oppenheimer, in 1935. As he described it, the proton in the deuteron is repelled by the nucleus and tends to hang back. The neutron in the deuteron isn't affected by the nucleus at all and keeps on moving. As a result, it often happens that the proton and the neutron in the deuteron are split apart by the strain. The proton glances away while the neutron keeps on and may strike the nucleus. The neutron has only half the energy of the original deuteron, but this may still be much more than the energy possessed by ordinary neutrons. As it turns out, the deuteron is the most useful particle for the production of radioisotopes.

Neutron Reactions

Once people started working with neutron bombardment, it gradually became the most interesting type of "atom-smashing."

When a neutron is absorbed by an atomic nucleus, it doesn't change the atomic number of that atom. The atom therefore remains the same element it was before. However, its mass number is increased by 1.

THE DEUTERON

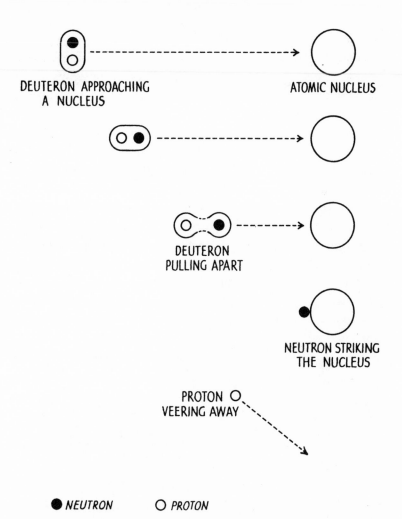

DEUTERON APPROACHING
A NUCLEUS

ATOMIC NUCLEUS

DEUTERON
PULLING APART

NEUTRON STRIKING
THE NUCLEUS

PROTON
VEERING AWAY

● NEUTRON   ○ PROTON

If the nucleus of hydrogen-1 absorbs a neutron, it becomes hydrogen-2. One stable isotope has been converted into another stable isotope, and nothing further happens. A neutron is absorbed, and that is all there is to it.

Often, though, a stable isotope is converted into an unstable isotope as a result of neutron absorption. For instance, indium-115 (stable) is converted into indium-116 (unstable). A new radiation gradually arises as the new isotope is formed; beta particles are thrown out, and tin-116 (stable) is formed.

Sometimes, when a neutron is absorbed by an atomic nucleus, another particle is thrown out at once. You may remember, from an earlier chapter, that when nitrogen-14 absorbs a neutron, it immediately throws out a proton. The resulting atom (with one extra neutron and one less proton) is carbon-14. There are also nuclear reactions in which neutrons are absorbed and electrons or alpha particles are immediately given off.

A particularly interesting type of neutron reaction is one in which a nucleus absorbs a neutron and then, as a result, immediately throws out a neutron. This nucleus isn't changed at all. It has taken a step forward, so to speak, followed by a step backward, and ends up where it began.

You may wonder why this forward-and-back action should be interesting. Well, when the bombarding neutrons are particularly energetic, two neutrons (or even more) are sometimes thrown out. Now the nucleus has taken a step forward and followed that by two (or more) steps backward.

An example of this kind of nuclear reaction occurs in the bombardment of carbon-12 (a stable isotope which makes up 99 percent of all carbon atoms) with neutrons. One neutron is taken up, and, if the neutron is energetic enough, two neutrons are thrown out. The net result is that the carbon has one less neutron than it started with. It is now carbon-11, which is unstable and gives off positrons. (The half-life of carbon-11 is about twenty minutes.)

Now observe that for the first time there's a chance of making a profit. Suppose you can bombard some kind of atom with a neutron hard enough to make it give up two neutrons. Suppose each of those two neutrons hits an atomic nucleus hard enough to make each nucleus give up two neutrons. That's four neutrons altogether. The four neutrons may then hit four nuclei to produce a total of eight. The eight can produce sixteen, then thirty-two, then sixty-four, and so on — all starting with a single neutron!

Suppose it takes one second to break up a nucleus. At the end of one second, two neutrons are produced. At the end of two seconds, four neutrons are produced. At the end of three seconds we have eight. At the end of four seconds we have sixteen. At the end of thirty seconds, one billion neutrons are produced! (If you can't believe this, work it out for yourself. Double the number of neutrons every second for thirty seconds on paper, and see what the final figure is.)

A nucleus actually breaks down under bombardment in only a millionth of a second. The neutrons therefore build up in number with incredible speed. Starting with one neutron, millions, billions, trillions of atoms could be breaking down in a trifling fraction of a second.

A nuclear reaction such as this, in which steps follow one another like links in a chain, is called a *chain reaction*.

Even if each breakdown of a nucleus liberated only the tiniest bit of energy, so many nuclei would be breaking down after a fraction of a second that the energy produced would be enormous. And it all starts with a single neutron. There's atomic energy for you!

As early as 1934, a Hungarian physicist, Leo Szilard, who was then working in England, considered the possibility of such a chain reaction. He even applied for a patent for it but kept his thoughts secret, for he foresaw great dangers.

There's only one trouble. You must find a nuclear reaction which liberates more neutrons than it uses up and in which the liberated neutrons are energetic enough to keep the reaction going. Usually you need very energetic neutrons to start such a reaction, and the neutrons produced by a nucleus are less energetic than the ones it absorbs. The carbon-12 reaction with neutrons won't do; the neutrons produced aren't active enough. Most chain reactions, even when theoretically possible, don't work. Even after you start them, they peter out, like a match flame in a high wind.

In 1939, however, a new kind of nuclear reaction was discovered that almost at once changed the course of human history.

Fission!

Although people talk about "atom-smashing," no atoms were knowingly "smashed" by man before 1939. The most man was able to do was to knock out one, two, maybe four particles. The atom was only being chipped away; it wasn't being "smashed."

This was changed by Enrico Fermi (who had named the neutrino and worked out its behavior). He was the first physicist to begin working with the bombardment of neutrons. He found that if the neutrons were slowed down, they were absorbed even more efficiently, and brought about

CHAIN REACTION

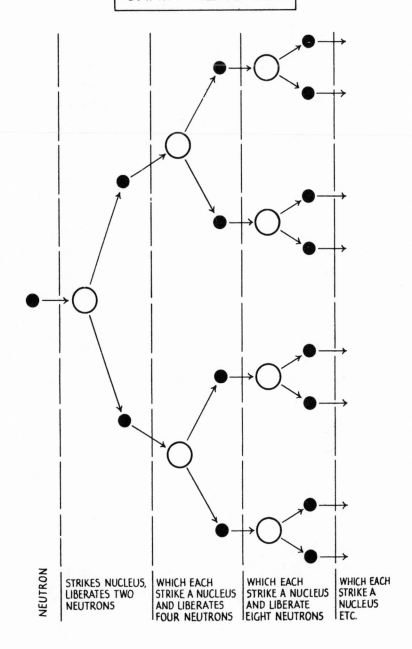

NEUTRON | STRIKES NUCLEUS, LIBERATES TWO NEUTRONS | WHICH EACH STRIKE A NUCLEUS AND LIBERATES FOUR NEUTRONS | WHICH EACH STRIKE A NUCLEUS AND LIBERATE EIGHT NEUTRONS | WHICH EACH STRIKE A NUCLEUS ETC.

nuclear reactions more easily. (He received a Nobel Prize in 1938 for his work with neutrons.)

In particular, he wanted to bombard uranium with neutrons. A neutron, when absorbed, produced a higher isotope of the element that was being bombarded, and this isotope often gave off a beta particle to become an element one higher in atomic number. If uranium (atomic number 92) did that, it would become element 93, and Fermi thought it would be interesting to bring that about.

For a while he thought he had succeeded (and indeed he had, but it was not until 1940 that McMillan and Abelson discovered element 93 — neptunium — in uranium that had been bombarded by neutrons). Unfortunately so many other things went on in the uranium that Fermi could get no clear answer as to what had happened.

Other scientists tackled the problem and couldn't get the answer either. In particular, a German chemist, Otto Hahn, and his Austrian associate, Lise Meitner, were interested. In 1938, Hahn felt he had solved the puzzle. When a uranium atom absorbed a neutron, it sometimes split in half. This was so odd that Hahn was actually afraid to say anything about it. He felt he couldn't be right and that he would simply be laughed at.

By that time, however, Germany under Hitler had marched into Austria, and the Austrian, Lise Meitner, came under Nazi law. Since she was Jewish, she had to leave the country, and escaped to Sweden. While in Sweden, she considered Hahn's discovery and decided that she would publish it. She did so, in January, 1939.

The Danish scientist, Niels Bohr, was just on the point of leaving for the United States to attend a conference, and Miss Meitner told him what she was doing. Bohr took the news with him and spread it among the American scientists he was meeting.

As soon as the theory became known, American scientists rushed to their laboratories, repeated the experiments, and at once Miss Meitner was shown to be right. As a result, Hahn received a Nobel Prize in 1944.

And so a new kind of nuclear reaction had been discovered, one in which a nucleus is not merely chipped, but is actually split. This kind of reaction is called *fission*.

The Special Case of Uranium-235

It turned out that many heavy nuclei, in addition to uranium, can split, or undergo fission, once they are struck with sub-atomic particles (usually neutrons) in the proper way. For most of them, however, the neutrons must be travelling at considerable speed. Most uranium atoms are no exception. Uranium-238, which makes up 993 atoms out of every thousand of that element, requires fast neutrons before it will split.

Uranium-235 is different. It is a special case as Bohr was the first to point out. A uranium-235 nucleus will split even when it is hit by a slow neutron, one which is just drifting along.

We might compare the two uranium isotopes to more ordinary things to see what this means. Uranium-238 (and most other heavy nuclei) is like a piece of hard wood you are trying to set on fire. An ordinary match won't do the trick. You need a bigger and hotter flame to get the log burning. Uranium-235, however, acts like paper. A touch of the match, and off it goes, burning.

Uranium-235, in fact, doesn't have to wait for man's help to undergo fission. There are free neutrons all around us at all times. They are produced when cosmic rays strike atoms in the atmosphere and elsewhere. These neutrons aren't very plentiful, really, but they are there. One of them, occasionally, may strike a uranium-235 nucleus and cause it to split. Or, once in a long while, a uranium-235 atom may undergo fission without any encouragement at all. This is called *spontaneous fission.*

Uranium-235 has a half-life for spontaneous fission, just as it has one for its ordinary production of alpha particles. The half-life for spontaneous fission is extremely long, however — millions of billions of years. For every uranium-235 nucleus that undergoes spontaneous fission, a million or more break down in the usual way by throwing out an alpha particle.

The new transuranium elements undergo spontaneous fission more easily. The fission half-life of plutonium-236 is 3½ billion years; of curium-240, 20,000 years; of californium-

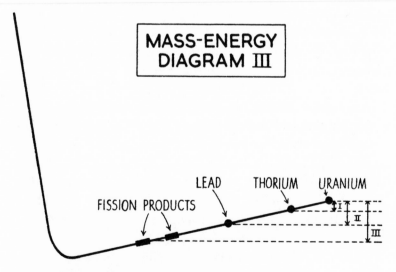

MASS-ENERGY DIAGRAM III

FISSION PRODUCTS   LEAD   THORIUM   URANIUM

I   *ENERGY DEVELOPED IN PRODUCTION OF ALPHA PARTICLES*
II *ENERGY DEVELOPED IN URANIUM BREAKDOWN TO LEAD*
III *ENERGY DEVELOPED IN URANIUM FISSION*

252, 100 years; and of fermium-254, only seven months. The very massive atoms, in fact, are likely to break down by spontaneous fission more quickly than in any other fashion.

If only scientists had thought of looking for it, they would probably have detected this spontaneous fission years ago, since the energy liberated is much greater in fission than in production of alpha particles. You will see why this is so in a moment.

When uranium-235 nuclei undergo fission, they don't split exactly in half. They don't even always split in exactly the same way. A variety of products are formed. In fact, thirty-four different elements have been detected among the products of fission. Most of the nuclei formed, however, have mass numbers from 85 to 104 and from 130 to 149.

Now look at Mass Energy Diagram III (page 176). When a uranium-235 atom loses an alpha particle, the mass number drops to only 231. When it breaks down all the way to lead, the mass number drops to 207. When fission takes place, however, the mass number drops at least to 149 and maybe all the way to 85. See how much more energy is produced in fission than in ordinary radioactivity. The energy is several times as great for each nucleus.

Here was a source of atomic energy better than any which science had known of before 1939. In a very few years, this energy was put to amazing use.

ATOMIC DANGERS

The Chain Reaction That Could Work

Every uranium-235 nucleus that undergoes fission can produce as much as 7,000 times the energy contained in the neutron that makes it break in two. That would seem like profit enough as far as energy is concerned, but it is only the beginning.

You may remember (see page 36) that atomic nuclei of small mass number are stable when they contain equal numbers of protons and neutrons. As nuclei grow more massive they need extra neutrons and the more massive they are, the more extra neutrons they need.

When the massive uranium-235 nucleus breaks into two middle-sized pieces, the two pieces don't need all those extra neutrons. There are now neutrons to spare. When the uranium-235 nucleus breaks up, not only are two smaller nuclei formed, but also two or three free neutrons!

Now we have what we need for the chain reaction described in the previous chapter. One uranium-235 nucleus splits up as a neutron hits it. Two or three new neutrons are formed. If these hit two or three other uranium-235 nuclei, these split up, too, and anywhere from four to nine neutrons are formed. In no time at all, every uranium-235 nucleus in sight is breaking up.

For the investment of a single neutron you get not only 7,000 times as much energy from one shattered uranium-235 nucleus. You get trillions upon trillions of times as much energy when all the nuclei start to go.

But will such a chain really carry on? Are the neutrons formed energetic enough to keep it going? That is the problem, you remember, in other neutron-releasing nuclear reactions.

The beauty of uranium-235 fission is that you don't need energetic neutrons to keep it going. Slow neutrons with very little energy will do the trick. In fact, slow neutrons are better for the purpose than fast neutrons.

As soon as fission was discovered, scientists at once realized the possibilities of this chain reaction. They foresaw the floods of energy that could be made available — unbelievable quantities of it.

There was still a question whether certain practical difficulties could be avoided or overcome. Ordinarily these difficulties might have held up scientists a long time. A great deal of work and money was necessary, you see, and nobody could guarantee in advance that things would work out. It was certain that the fission chain reaction could work in theory. But suppose the practical difficulties turned out to be too great or complicated. Private individuals or institutions might not have wished to invest enough money in what might, after all, turn out to be a wild-goose chase.

But suppose a government was involved in the work; suppose the American government was involved in it, the richest government in the world.

Leo Szilard was thinking about this. He was in the United States now, having fled Europe to escape Nazi tyranny. He realized that war was about to break out and that mankind faced an enormous danger. He had already been working on chain reactions (see page 171), and he saw that uranium fission could be made to produce a practical one.

What if Hitler and his Nazis obtained the secret of a practical fission chain reaction first? They might conquer the world. Somehow the American government had to be made interested in what might seem a "far-out" project. He interested two other physicists of Hungarian birth, who now lived in America, in the project. They were Eugene Paul Wigner (who received a Nobel Prize in 1963 for his theoretical work on atomic structure) and Edward Teller.

All three went to Albert Einstein, who also lived in the United States now. Einstein was the most famous scientist in the world, and perhaps people would listen to him. They persuaded Einstein (who dreaded the results of working with fission) to write a letter to President Franklin D. Roosevelt.

In 1941, Roosevelt was persuaded, and he agreed to start a large research program aimed at developing a war weapon based on uranium fission. The program was to be called by the name *Manhattan Engineering District* so that no one would guess what it was about, but it is popularly called the "Manhattan Project."

The order establishing the research program was signed on December 6, 1941. The next day Japan bombed Pearl

Harbor, and the United States was at war. Had Roosevelt waited one more day, the order might have slipped his mind in the excitement of events.

Altogether, the United States spent two billion dollars meeting and overcoming the difficulties that had to be beaten before the war weapon· could be made practical.

Some of the Difficulties

In the first place, if we are to have a chain reaction, the neutrons produced in uranium-235 fission must hit other uranium-235 nuclei. But suppose the neutrons hit other types of nuclei instead. If they do, they may be absorbed. After that nothing may happen, or else some ordinary nuclear transformation that doesn't produce neutrons may take place. In either case, the neutrons are gone, and the chain reaction fizzles out.

It is simple enough to purify uranium and make sure that only uranium atoms are in the neighborhood of splitting uranium-235 nuclei. Unfortunately, that isn't enough. Even in pure uranium, most of the nuclei are uranium-238, the wrong kind. They won't undergo fission except under special circumstances. Only seven uranium atoms out of a thousand are uranium-235. This is not enough for good results. Somehow the number of uranium-235 atoms must be increased and the number of uranium-238 atoms decreased.

This means that it is necessary to separate uranium-235 and uranium-238. In an earlier chapter we explained why the separation of isotopes is very difficult. Now scientists were faced with the necessity of doing it on a large scale. They needed pounds and pounds of fairly pure uranium-235. That was one of the chief difficulties.

The separation of the uranium isotopes was tried in half a dozen ways. One was by using special mass spectrographs.

Another method began by combining uranium with a gas called fluorine to form a gas called *uranium hexafluoride*. Some of the uranium hexafluoride molecules contained uranium-238 and some contained uranium-235. The molecules with uranium-235 were about 1 percent lighter than the others. When the gas was forced through a number of partitions containing tiny holes, the molecules containing uranium-235 were just a trifle nimbler than the rest and managed to get through the holes a trifle faster. The first portions of the gas getting through the final barrier contained only uranium-235 molecules.

In a surprisingly short time, scientists were preparing uranium-235 in quantities that would have seemed completely impossible only a few years earlier.

Enriched uranium (uranium containing greater amounts of uranium-235 than do natural samples) could now be used to support a chain reaction. Remember, it isn't necessary for every single neutron formed through fission to hit another uranium-235 nucleus. It is only necessary for at least one neutron to hit a uranium-235 nucleus for every uranium-235 nucleus that undergoes fission. In that way the number of nuclei undergoing fission in each instant stays the same or even increases. So does the flow of energy.

You may wonder why scientists needed so much uranium-235. Couldn't a small piece, a tiny piece, do for the first experiment? The answer is no. That is another difficulty.

Suppose you do start with a small piece of uranium, say an ounce. If fission begins within that piece because a neutron hits a uranium-235 nucleus, then neutrons start flying about, and other nuclei ought to be split. But a neutron may hit a number of nuclei and bounce off harmlessly before striking one sufficiently dead-center to be absorbed and start fission going. If the piece of uranium is small, many of the neutrons may find themselves outside the piece altogether before such a dead-center hit has been made; so many neutrons are lost to the atmosphere that those left are not enough to keep the chain reaction going. It dies down and out.

If you start with a larger piece of uranium, there is a bigger chance that the neutrons will make a sufficiently direct hit upon a uranium-235 nucleus before finding their way out of the piece altogether. As the piece grows larger, so does the chance. When the quantity of uranium reaches a certain amount, enough neutrons hit nuclei to allow the chain reaction to proceed and become self-supporting. The smallest quantity of uranium that will let this happen is called the *critical size*.

The Atomic Bomb

Now suppose you had two masses of enriched uranium, each of which was a little smaller than the critical size. Fission of atoms would be going on here and there in each of the masses. No chain reaction would be set up, however. Most of the neutrons resulting from fission would escape from the uranium into the air.

But what if the two masses, at a certain moment, were pushed forcibly together? They would become a single mass which would be greater than the critical size.

What would happen the instant the two masses plunged together? Somewhere within that mass uranium-235 atoms would be undergoing fission. (You wouldn't have to bombard the mass with neutrons. Spontaneous fission would be going on, you know.) But now the chain reaction would be set up. The neutrons being formed by fission would hit other uranium-235 nuclei before working their way out of the uranium. More neutrons would form, and still more.

The chain reaction wouldn't just hold its own, either. The

number of nuclei that were splitting would increase every instant. In a few millionths of a second, uncounted numbers of uranium-235 nuclei would break up. Each individual fission would contribute a small bit of energy. Before very long you would expect the uranium to melt, boil, and vaporize away. That would end the chain reaction.

Before the reaction ended, the energy produced would have mounted with unbelievable quickness. Before the uranium could boil or blow apart, a brilliant fireball, as bright as the sun or brighter, would have been formed. X-rays and gamma rays would have poured out, speeding particles would have sprayed in all directions, and the temperature within the fireball would have risen as high as ten million degrees.

That is what scientists guessed would happen. Naturally, they were eager to test a piece of enriched uranium and see if their guess was correct.

Ordinarily, if a new explosive is being tested, the thing to do is to take a small quantity of it and try a very small explosion. This was impossible in the case of uranium-235. You had to have the critical size, or it would not explode. If you did use the critical size, you would have a tremendous explosion. It was all or nothing.

The scientists had no choice. They tried all. On July 16, 1945, in Alamogordo, New Mexico, the first *atomic bomb* in history was exploded, with terrifying results. It was like the explosion of thousands of tons of TNT all at once.

Since then most of us have seen pictures of such explosions, in newspapers and magazines, in newsreels and on television. We all know what a frightening sight it is.

By July, 1945, the war in Europe was over and Nazi Germany was completely destroyed. They, too, had been doing research on atomic weapons, under Heisenberg, but they hadn't made much progress. Japan was still in the war, though. It was already on its last legs, but the American government decided to hurry matters along (against the advice of a number of scientists who were concerned at the deadly power of the weapon they had invented).

Two more bombs were prepared, and the very next month, in August, 1945, they were exploded over the cities of Hiroshima and Nagasaki in Japan. The damage they did was tragic. The centers of the cities were destroyed, and over a hundred thousand people were killed or hurt. It was the last straw for an already badly damaged Japan. Within two weeks, Japan surrendered, and World War II was over.

Since then, the United States has exploded hundreds of atomic bombs, but only for testing purposes. No atomic bomb has been used in warfare since, and all mankind prays no such bomb ever will be.

Nor is the United States the only nation to possess such bombs. For four years we had an atomic monopoly, then, in 1949, the Soviet Union exploded an atomic bomb. In 1952, Great Britain exploded a test bomb, and in 1960 France did. In 1964, the Chinese People's Republic ("Communist China") exploded one and became the fifth member of the "atomic club."

While this was going on, however, bombs much more powerful than the type exploded over Hiroshima and Nagasaki were produced.

The Hydrogen Bomb

Suppose we look at Mass-Energy Diagram IV (page 186). The first atomic bombs (or *A-bombs*, as they are often called) involved uranium. Uranium is at the right-hand end of the diagram.

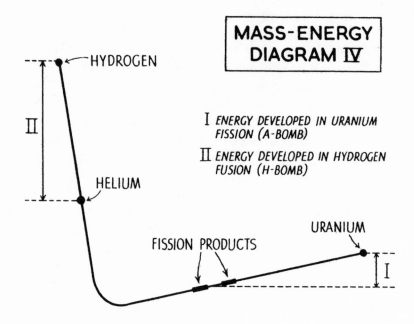

HYDROGEN

II

I ENERGY DEVELOPED IN URANIUM
FISSION (A-BOMB)

II ENERGY DEVELOPED IN HYDROGEN
FUSION (H-BOMB)

HELIUM

URANIUM

FISSION PRODUCTS

I

That end of the diagram is the flattened end. In sliding down the energy hill from uranium to middle-sized nuclei, the atoms don't lose very much mass. It is enough to give us our powerful atomic bombs, to be sure, but compare it with the other end of the diagram.

Suppose we convert hydrogen to helium, for instance. Notice how much more mass would disappear, how much more energy would be produced. In an ordinary A-bomb, about one-tenth of one percent of the mass is converted to energy. In a hydrogen bomb, seven-tenths of one percent of the mass is converted to energy.

Scientists, from 1949 onward, worked seriously on methods for releasing this energy.

To change hydrogen to helium, something new had to enter the picture. Back on page 128 I described one way of bringing about nuclear reactions. This method involves the use of energetic sub-atomic particles as atomic bullets. I have been talking about that method ever since. Now it is time to describe a second method.

That second method is heat! Nothing more than that.

This may puzzle you. You may remember that I also said, earlier in the book, that radioactivity and other nuclear reactions aren't affected by heat. What I meant was that they are not affected by the amount of heat that mankind could produce before 1945.

Radioactive substances could be heated white-hot, but what of that? The temperature might go up a few thousands of degrees. This would cause the atoms to bounce together much harder than usual, true. Still, the electron shells would protect the nuclei so that nuclear reactions wouldn't be affected.

But what if, instead of temperatures of a few thousand degrees, mankind could produce temperatures of millions of degrees?

At such unbelievable temperatures, the electrons would be stripped away from the nuclei, the nuclei would be smashed together, and nuclear reactions would result. Reactions such as these, brought about by extreme heat, are called *thermonuclear reactions*. (The "thermo" prefix comes from a Greek word meaning "heat," and we are familiar with it in words such as "thermometer" and "thermos bottle.")

Such heat could be supplied by an exploding A-bomb.

New bombs were therefore manufactured in which an ordinary uranium-235 explosion just acted as a trigger. The A-bomb set off a much huger and more tremendous reaction in which hydrogen atoms were converted into helium atoms.

These new bombs were called *hydrogen bombs* (or *H-bombs*). Ordinary uranium-235 bombs are examples of a *fission bomb.* A hydrogen bomb, in which hydrogen atoms are fused together to form helium, is a *fusion bomb.*

Actually, heavy hydrogen isotopes, deuterium and tritium, are used in H-bombs, rather than ordinary hydrogen. The heavy hydrogen is combined with lithium to form lithium hydride, a solid substance. The lithium and heavy hydrogen can combine to form helium, liberating enormous quantities of energy.

The exact method used to bring about the H-bomb explosion is a military secret, of course. It seems, though, that the key to the method was first proposed by Edward Teller (one of the three physicists who persuaded Einstein to write his letter to President Roosevelt). For this reason, Teller is widely known as the "father of the H-bomb."

Experimental hydrogen bombs have been exploded by the United States, Great Britain, and the Soviet Union. The A-bomb exploded at Hiroshima was as powerful as twenty thousand tons of TNT, but H-bombs can be as powerful as fifty million tons of TNT. A single hydrogen bomb can destroy even the largest city almost completely.

The sun (and the average star) obtains its energy from thermonuclear reactions. The temperature in the interior of the sun is in the millions of degrees. There the sun's hydrogen (which makes up 85 percent of its whole volume) is being converted into helium. In this way, the sun loses mass and releases energy. All of us, therefore, are living in the light and warmth of a huge hydrogen bomb, 860,000 miles across and 93,000,000 miles away, which is in a state of continuous explosion.

The Three Effects

A single bomb exploding with a force equal to millions of tons of TNT is certainly a terrible thing for the human

ATOMIC POWER

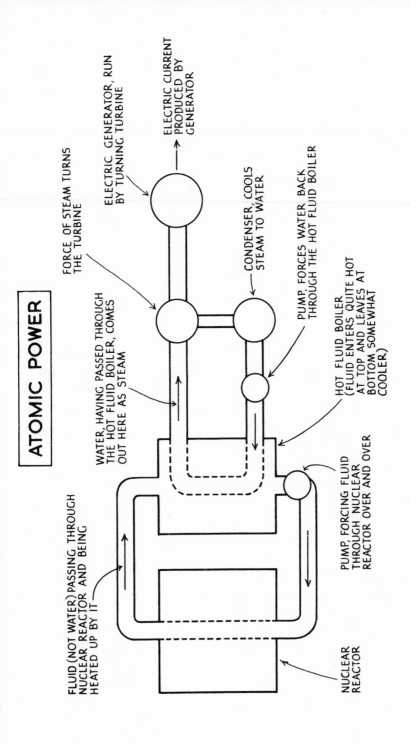

FORCE OF STEAM TURNS THE TURBINE

ELECTRIC GENERATOR, RUN BY TURNING TURBINE

ELECTRIC CURRENT PRODUCED BY GENERATOR

WATER, HAVING PASSED THROUGH THE HOT FLUID BOILER, COMES OUT HERE AS STEAM

CONDENSER, COOLS STEAM TO WATER

PUMP, FORCES WATER BACK THROUGH THE HOT FLUID BOILER

HOT FLUID BOILER (FLUID ENTERS QUITE HOT AT TOP AND LEAVES AT BOTTOM SOMEWHAT COOLER)

FLUID (NOT WATER) PASSING THROUGH NUCLEAR REACTOR AND BEING HEATED UP BY IT

PUMP, FORCING FLUID THROUGH NUCLEAR REACTOR OVER AND OVER

NUCLEAR REACTOR

race to have to face. Unfortunately, a nuclear explosion would be more dangerous than a TNT explosion even if the TNT explosion were just as powerful.

A-bombs and H-bombs do damage in three ways.

The first damaging effect is that of *blast*. The explosion forces air outward in gigantic "shock waves." The earth, too, trembles as a result of the concussion. The combination of wind and earthquake will knock down houses and structures for miles around. It will blast man and his works to smithereens.

The second damaging effect is that of *heat*. The tremendous temperatures produced will set widespread fires. This will add to the havoc of the blast and may spread its effects even further.

These two effects, however, are similar to those of ordinary explosives. The blast and heat of atomic bombs are much worse than those of ordinary chemical explosives, but they are things with which mankind is familiar.

The third effect, however, is one that is not produced by ordinary explosives, and it is the worst danger of all. This third effect is *energetic radiation*.

Some of the radiation of a nuclear explosion appears in the form of blinding light, but there is also an invisible radiation in the form of x-rays and gamma rays. These invisible and very energetic radiations can damage any living tissue that they penetrate. The most important compounds in living tissue are made up of huge molecules consisting of thousands or even millions of atoms. Some of these huge molecules are very fragile and finely balanced. When a gamma ray goes bowling into such a molecule, it may knock pieces off, or it may bring the whole thing down like a house of blocks.

People who are exposed to such energetic radiation may fall victim to *radiation sickness*. If the exposure is great enough, radiation sickness can be fatal. In fact, in the early days of research in x-rays and radioactivity, a number of scientists died of the effects. It was a number of years before people learned to take proper precautions.

Nowadays, the quantity of x-rays (and gamma rays too) is measured by the number of ions produced. The quantity of radiation is then said to be so many *roentgens*, where one roentgen is the amount of x-rays or gamma rays which is sufficient to form two billion ions in each cubic centimeter of air. Scientists have tried to work out how many roentgens of such radiations human beings can absorb without damage. A limit of 0.3 roentgens per week is one limit that has been set by some authorities.

Fallout — The Worst Danger Of All

The gamma rays formed at the instant of atomic explosion are bad enough, but they pass as the explosion dies down and are gone in a minute or less. Worse, much worse, are the products of the nuclear reactions that have gone on during the explosion. These products remain behind after the explosion is over. They are highly radioactive and keep breaking down and releasing high-energy radiations.

The radioactive products, combined with pulverized and vaporized soil, spread out in the upper atmosphere (you have all seen pictures, probably, of the now familiar "mushroom cloud"). The winds carry these radioactive isotopes for hundreds and thousands of miles.

That cloud of radioactive particles being moved by the wind is called the *fallout* of the explosion. It is deadly. The fallout from a large hydrogen bomb explosion, spreading radiation sickness or even death for hundreds of miles down-wind, may cause many more casualties than the explosion itself.

It is possible to add certain elements to an H-bomb so that particularly dangerous isotopes will be formed. Then the fallout becomes more deadly than ever. One H-bomb can produce enough fallout to be fatal over several thousand square miles. The fallout from one of America's experimental H-bombs actually struck a Japanese fishing boat and the fishermen were attacked by radiation sickness. One died—the first man in history to be killed by an H-bomb.

The most dangerous isotope formed in the fallout is *strontium-90*. Strontium is an element very similar to the calcium in our bones. Our body picks up strontium-90 and, mistaking it for calcium, stores it in the bones, where it stays for a very long time. Since the half-life of strontium is 28 years, strontium-90 stays radioactive for a lifetime. Radioactive substances in our bones could cause serious diseases. So far the amount we have collected is not enough to be dangerous, but we all do have some of it in our bones now, whereas fifteen years ago nobody did because strontium-90 did not exist then.

Scientists grew so aware of the fallout danger that even test explosions in Nevada were performed with tremendous care. They were postponed from day to day until weather conditions were just right to ensure that the fallout would travel over uninhabited regions and would do no damage.

Eventually, the radioactive cloud gradually spreads thin and is blown apart by the wind. Even then, radioactive atoms find their way to all parts of the atmosphere, and to the oceans, too, when the blast is near water. The amount of radioactivity that reaches any particular place isn't large, but it is enough to be detected by delicate instruments.

Every scientific instrument designed to detect radio-activity is always measuring a small amount of it. This is due to the naturally radioactive atoms that are always present in tiny quantities in the soil and atmosphere. The action of cosmic rays also contributes. At sea level, cosmic rays contribute less than half of the total. At elevations of a mile, where there is less atmosphere above one to absorb the cosmic rays, the cosmic ray contribution rises to slightly more than half. The small quantity of radioactivity that is always present is known as *background radiation*.

Whenever an atomic explosion takes place in the atmosphere anywhere in the world, the background radiation everywhere else in the world goes up for some hours or days afterward, the time depending on how distant the explosion was. For that reason, no atomic explosion of this sort can be kept secret. When the Soviet Union exploded its first atomic bomb in September, 1949, American scientists knew very shortly after that it had happened.

There is some question whether repeated atomic explosions in the form of tests and experiments may not be increasing the background radiation permanently. The increase so far noticed is not enough, nowhere near enough, to damage living tissue at once. However, some scientists wonder if continued exposure may not cause tiny bits of unnoticed damage that can be inherited by our children. Through the years and generations, then, the whole human race may be weakened by atomic experiments. This is a problem to be taken very seriously. We don't have the answer yet, but many keen minds are doing their best to find it.

One of the scientists most concerned about the danger of fallout from nuclear testing is the American chemist, Linus Pauling. He has spoken without ceasing about that danger and called tirelessly for an end to the testing of atomic bombs. As a result he was awarded the Nobel Prize for peace in 1963. Back in 1954, he had already received a Nobel Prize for his chemical work. This meant that Pauling and Madame Curie were the only two people ever to earn two Nobel Prizes.

Such was the pressure against atomic bomb testing that in 1958, the United States, Great Britain, and the Soviet Union joined in a "gentleman's agreement" to test no more bombs. In 1961, the Soviet Union broke that agreement, and a new round of testing began. In 1963, though, a formal treaty was signed by those three nations, banning all tests save those conducted underground. Underground explosions, properly done, produce no fallout. It is to be hoped that this will not be broken.

Two members of the "atomic club" have not signed this *test-ban treaty*. They are France and China. They, however, possess the earlier fission bombs only and not the dreadful H-bombs. At least, they don't have the H-bombs yet.

The United States and the Soviet Union have exploded nuclear bombs in outer space in the past, but that doesn't seem to be a good idea, either. They don't produce fallout but they affect the structure of the upper atmosphere and charged particles outside the atmosphere, and scientists object that this may get in the way of research.

Both the United States and the Soviet Union have carried out explosions of small atomic bombs underground since the test-ban treaty. This seems to be fairly safe.

ATOMIC HOPE

The Delicate Balance

It would be a pity if man's growing knowledge of the atom resulted only in A-bombs and H-bombs. Such terrible weapons are not pleasant to consider, least of all by the scientists whose work made them possible.

But good things also result from atomic research, as we shall see.

There is a happy medium in uranium fission. If a lump of uranium is below the critical size, fission may start, but the chain reaction will die out. If the uranium is above the critical size, there is an almost instantaneous and fearful explosion. But what if a lump of uranium is at exactly the critical size? Then fission can start and maintain itself! The chain reaction will proceed at an exactly even rate.

That should make everything simple, you may think. It should be easy to make a lump of uranium of exactly the right size, and then everything would take care of itself. But the problem is not so simple as that. The trouble is that the critical size is not always exactly the same; it depends, for example, on the speed of the neutrons that are moving through the lump.

Fast neutrons, it is easy to see, have a better chance than slow neutrons of escaping from the lump without hitting and splitting a uranium-235 atom. If we use fast neutrons, many of them will escape from the lump without making a hit; there may not be enough hits to keep the chain reaction going. If we make our lump bigger, however, the fast neutrons will have farther to go to escape from the lump and thus they will have a better chance of hitting uranium-235 atoms on the way; there may now be enough hits to keep the chain reaction going. The use of fast neutrons, we see, has increased the critical size. In the opposite way the use of slow neutrons decreases the critical size.

It would not be practical to change the size of a lump of uranium while fission was going on. There is a better way out. Instead of changing the actual size to fit the critical size, we change the critical size to fit the actual size. We keep the critical size the same as the actual size by regulating the number and speed of the neutrons. First we build in a "throttle" that will speed up fission by supplying plenty of neutrons of the right kind; then we build in a "brake" that will slow down fission by removing some of those neutrons. By adjusting the "brake" we can keep the fission going at an even rate.

First the "throttle": Slow neutrons, you remember from an earlier chapter, are more efficient than fast neutrons in splitting uranium-235 atoms. We want plenty of slow neutrons. How do we get them? By slowing down fast neutrons.

Fission starts off, you remember, either by a spontaneous breakdown of the uranium-235 nucleus or by the action of slow neutrons in the atmosphere (which were put there by the action of cosmic rays). Once fission starts, however, the neutrons produced by the splitting uranium-235 atoms are fast neutrons. These must be slowed up.

To understand how this is done, you must think for a moment of a billiard ball. If a billiard ball is struck by a cue and goes speeding into a cannon ball, it simply bounces off. The cannon ball is hardly affected. The billiard ball changes its direction but moves just as fast as before.

Suppose, however, that the billiard ball strikes, instead of a cannon ball, another billiard ball. It still bounces off, but this time the second billiard ball moves also. The speed gained by the second billiard ball must be lost by the first one. (You can't get something for nothing). Neither ball moves as rapidly as the first one did originally.

If the billiard ball speeds into a group of fifteen billiard balls (a ball does that at the start of certain games), all the billiard balls move, but quite slowly. The original billiard ball slows to a crawl, also. The original motion has been spread among sixteen different balls, and none of them, not even the original ball, has more than a small share of it.

We must do something like that to the fast neutrons. We must add to the uranium a substance whose atoms will slow up fast neutrons without absorbing them. If the substance is made up of massive atoms, the neutrons will just bounce off and be as fast or nearly as fast as ever, like the billiard ball that hit the cannon ball. If the substance is made up of light atoms, not very much heavier than the neutron, the neutron will slow up, like the billiard ball hitting other billiard balls.

Light atoms with small cross-sections will slow up fast neutrons without absorbing them. A substance made up of such atoms is called a *moderator*. Some of the moderators used in fission work are deuterium, beryllium, and carbon. The deuterium atom is only twice as heavy as the neutron, beryllium is nine times as heavy, and carbon is twelve times as heavy.

A fast neutron has to bounce off about two hundred carbon atoms before it slows down enough. The deuterium atom, being smaller than the carbon atom, works better. A fast neutron need bounce off only about fifty deuterium atoms. (Deuterium is used in the form of heavy water, and carbon is used in the form of graphite, which is the substance found in ordinary pencils.)

So much for the "throttle." The use of moderators, by increasing the supply of slow neutrons, helps to keep fission from dying down. Now how do we avoid the opposite danger? How do we keep fission from building up to an explosion? What is our "brake"?

For this purpose the element *cadmium* comes in handy. At least one of the stable cadmium isotopes has a very high cross-section. That is, it can absorb a neutron very easily and become another stable cadmium isotope with a mass number one unit higher.

If cadmium is present in the splitting uranium, a number of the neutrons formed are lost to the cadmium. If enough cadmium is present, not enough neutrons are left to keep the chain reaction going. Putting cadmium in the uranium is like putting water on a fire. Boron is another element that will work that way.

Birth of the Atomic Age

The first attempt at getting fission to set up a chain

reaction was at Columbia University in July, 1941. The reaction didn't become self-supporting because the uranium compound used wasn't pure enough. Too many neutrons were lost when the atoms of the impurities absorbed them without fissioning.

The first fission reaction system that worked and was self-supporting was built at the University of Chicago. Under the leadership of the Italian scientist Fermi, who had emigrated to America, a huge cube of uranium and carbon was built under the stands of a football stadium. First there was a layer of uranium, then a layer of carbon, then another layer of uranium, then another of carbon, and so on. Because these layers were piled one on top of another, the structure was called an *atomic pile*. Not all fission reaction systems are built this way, however, and a better name is *nuclear reactor*.

At various places in the atomic pile there were holes into which long rods of cadmium could be fitted. When the reactor was finished, it was 30 feet wide, 32 feet long, and 12 feet high. It weighed 1,400 tons and contained 52 tons of uranium.

Now, then, fission starts in the uranium. The neutrons are slowed down by the carbon, and they are absorbed by the cadmium. When the cadmium rods are kept all the way in the holes, so many neutrons are absorbed that fission dies down. As the cadmium rods are slowly pulled out, less and less cadmium remains inside the pile to absorb neutrons. At a certain point, the cadmium remaining inside the pile will absorb just enough neutrons to prevent an explosion, but not quite enough to cause the fission reaction to die down. The fission reaction is exactly balanced.

At 3:45 P.M. on December 2, 1942, the cadmium rods were pulled out just enough to let this happen, and for the first time a self-supporting fission reaction was set up. That day and minute really mark the beginning of the "atomic age." The first atomic bomb wasn't exploded for another two and a half years, because it took that long to produce enough uranium-235. However, once a self-supporting fission reaction could be set up, a bomb was just a detail.

News of this success was announced by a telegram reading: "The Italian navigator has entered the new world." (The Italian navigator had once been Columbus, but now it was Fermi, who discovered not the new world of the Americas but of atomic power.) There came a questioning wire in return: "How were the natives?" and the answer was sent off at once, "Very friendly."

Well, time will tell how friendly they are. It depends entirely on how they are treated by us, the navigators who follow Fermi.

The cadmium absorbers make it possible to control the atomic pile. Automatic devices can push them in or out to decrease or increase the flow of neutrons as necessary.

Fortunately, some of the neutrons emitted by fissioning uranium-235 are delayed. This leaves just enough time for the cadmium rods to be adjusted in case the fission reaction starts building up. If it weren't for this delay, the moment a build-up of fissioning atoms above the point of exact balance started, there would be an explosion. There would be no time to stop it even by the fastest-working machinery.

The first atomic pile at the University of Chicago wasn't very practical. It didn't have any system for keeping the uranium cool as fission proceeded, and there wasn't any adequate system for protecting people from radiation. Since 1943, however, a number of nuclear reactors of better design have been built, not only in this country but also in other countries, such as the Soviet Union, Great Britain,

France, Canada, Norway, West Germany, Japan and others. By the 1960's, even under-developed nations such as India were in possession of nuclear reactors. (This, however, is not the same thing as possessing atomic bombs.) Altogether, some 400 nuclear reactors now exist, about three-fourths of them American.

Isotopes A-Plenty

One thing the nuclear reactors made possible was the manufacture of radioactive isotopes in quantity. Before reactors were constructed, radioactive isotopes could be formed only in small amounts by the use of cyclotrons and similar instruments.

Nuclear reactors supply atomic bullets (in the form of neutrons of all energies) in much greater quantity than any cyclotron man has ever built. It is only necessary to stick some substance into holes in the reactor, specially prepared for the purpose. The atoms in that substance are then bombarded by uncounted numbers of neutrons. When the substance is taken out, a good proportion of it has changed to new isotopes, usually radioactive.

Samples of material containing such isotopes are prepared in powdered form or in liquid form, and are shipped all over the world in shielded containers. The strength of radioactivity of such samples is measured in *curies,* named after Pierre and Marie Curie, of course. A material containing 1 curie of radioactivity is liberating 37 billion particles each second.

Actually, a curie is quite a lot of radioactivity. Many samples contain only a few *millicuries* or even *microcuries* of radioactivity. A millicurie is a thousandth of a curie and a microcurie is a millionth of a curie. Sometimes radioactivity is measured in *rutherfords* (after the scientist who first brought about a nuclear reaction). One rutherford is equal to 1/37 of a millicurie, and it indicates the breakdown of one million atoms per second.

Once World War II was over, the world of science and medicine found that isotopes were available in quantity and at reasonable prices. The price of carbon-14, for instance, is about thirty thousand times less than it would have been had the nuclear reactor not been developed.

There seems to be no limit to the usefulness of these isotopes. Chemists can detect such small quantities of radioactive substances that they can track down chemical substances as they never could before. For instance, they can get more accurate figures on exactly how much of certain insoluble chemicals goes into solution in water. They can also look at chemical reactions more closely and decide exactly which atoms shift from compound to compound and how. This is extremely important in chemical theory.

Radioactive isotopes can even be used to tell us facts about living organisms. As far back as 1923, for instance, the Hungarian chemist, Georg von Hevesy, watered plants with water containing small quantities of a radioactive isotope of lead. He could follow the radioactivity easily, and this helped him come to certain conclusions about how plants absorbed minerals.

This was a poor experiment, for lead is not a natural part of living tissue, and there was no way of telling whether the plants weren't behaving abnormally because they were dealing with an unnatural substance. However, von Hevesy's experiment pointed the way, for when radioactive isotopes were available for elements that occurred naturally in living tissue, chemists had an excellent tool with which to explore the chemistry of such tissue. (This branch of the science is called *biochemistry.*) Von Hevesy was eventually awarded a Nobel Prize in 1943 for his pioneer work.

The use of radioactive isotopes represented a wonderful break for biochemists. You see, in any living creature, such as a human being, thousands upon thousands of chemical reactions are all going on at the same time in all parts of the body. Naturally, chemists would like to know what these reactions are. If they knew and understood them all, a great many of the problems of health and disease, of life, aging, and death, might be on the way to solution. But how are all those reactions to be unravelled? Not only are they all going on at the same time, but there are different reactions in different parts of the body and different reactions at different times in the same part of the body.

It is like trying to watch a million television sets all at once, each one turned to a different channel, and all the programs changing constantly.

But now radioactive isotopes come to the rescue. The body can't tell the isotopes of an element apart. If some of the food we eat contains carbon-14 (radioactive) instead of carbon-12 (stable), the body treats the two exactly alike.

Scientists, however, can tell the difference because carbon-14 gives off beta particles and carbon-12 doesn't. By following the trail of beta particles they can trace any compound containing carbon-14 through any changes it may undergo. For this reason, carbon-14 and other isotopes useful in such studies are called *tracers.*

A rat, for instance, may be fed a certain chemical which contains one or more atoms of carbon-14. (Methods of manufacturing chemicals containing radioactive isotopes are now well developed.) After a while, the rat is painlessly killed, and its tissues are analyzed. A number of different chemicals may be found, all containing carbon-14. They are identified, and then it is known that each of them must have been formed from the original chemical that had been fed to the rat.

By many such experiments (using isotopes other than carbon-14, too) many of the reactions of living tissue have been put together. It is a little like solving a jigsaw puzzle. The most important reaction of all is being worked out in this way. This is the reaction by which plants trap the energy of the sun and turn carbon dioxide and water into food. The reaction, called *photosynthesis,* may some day be duplicated by man and, if so, it would revolutionize our food situation.

For his work in studying photosynthesis by means of tracers and working out the jigsaw puzzle of its reactions, the American chemist, Melvin Calvin, received a Nobel Prize in 1961.

Radioactive isotopes can sometimes be used in medical diagnosis or treatment. One case involves the element *iodine*. Iodine is necessary, in very small quantities, to the body's working. It performs its function in the body by becoming part of a hormone which controls the rate at which we produce energy from the food we eat. This hormone is formed in the *thyroid gland,* a small mass of tissue located in the throat near the Adam's apple.

Small quantities of iodine ions or compounds are absorbed from our food into the blood. The bloodstream carries them to the thyroid gland, where the iodine is "trapped." It accumulates there. A radioactive iodine isotope would be trapped by the thyroid as readily as a stable iodine isotope.

Sometimes, when the thyroid gland is diseased, attempts are made to allow radioactive iodine isotopes to get to the thyroid in the hope that their radiations will kill off the diseased parts of the gland.

Radioactive isotopes are useful in almost every branch of science and engineering. Special laboratories have been built to handle them (for radioactive isotopes can be dangerous to health). Special devices have been designed to enable men to work with them at a distance. Whole books can be written on the uses of radioactive isotopes that have been developed since the end of World War II. And new uses are continually being found.

A few examples: Cobalt-60 is a radioactive isotope which can be formed quite easily by bombarding cobalt-59 with neutrons. Cobalt-60 throws off beta particles and rather powerful gamma rays. Small "needles" of cobalt-60 can be surrounded by metal to stop the beta particles and then the gamma rays can be used to kill cancer cells within the body, if the cancer is in a place that can be reached by the needle. Cobalt-60 is much cheaper than the radium that was used earlier and much safer, too.

Radioactive isotopes can be used to learn some of the facts about friction. It is friction that causes the wearing away of moving parts in machinery, and this, of course, is a source of great loss to industries and factories. Friction can be studied in this way: a steel piston ring is bombarded by neutrons until radioactive isotopes have formed in it. A lubricated piston is then moved up and down outside the rings just as it is in an automobile engine. Some of the steel in the ring is rubbed away and enters the lubricating fluid. By measuring the radioactivity of the lubricating fluid, very delicate measurements can be made of the amount of metal lost, and how that amount varies as conditions are varied.

When alloys are bombarded by sub-atomic particles such as neutrons, some of the atoms in the alloy (which is a mixture of metals) are knocked out of place. This changes certain properties of the alloy, at least temporarily. There may come a time, then, when alloys can be made to order not just by mixing metals according to some proper recipe, but by adding, in addition, just the right dash of radiation.

Neutron bombardment can be used to identify elements, too. Suppose a sample of unknown material is bombarded by neutrons. The various elements in the sample will absorb the neutrons and form radioactive isotopes. Each radioactive isotope will give off gamma rays of a certain wavelength and with a certain intensity. From this, chemists can now tell the nature and amount of elements in tiny samples of material. This new technique is called *neutron activation analysis.*

Radiation may be used to sterilize insect pests and cut down their ability to multiply. It may also be used to bring about changes in plant characteristics, producing new varieties from generation to generation. Some of these new varieties may prove to be more resistant to cold, drought or plant diseases. These are some of the ways in which radiation can help agriculture and increase the world's food supply.

A very interesting application of isotopes involves carbon-14, which is produced naturally by the cosmic rays, as was described earlier in the book. There is carbon in all living tissues and naturally some of that carbon is carbon-14. A dead creature, however, stops incorporating carbon-14 into its body. The carbon-14 that already existed within it slowly breaks down. Scientists know exactly how fast the carbon-14 breaks down and by measuring how much is left, they can tell how long it has been since the object stopped living. Old wood in Egyptian tombs, scraps of cloth in old graves, and similar objects between 1,000 and 30,000 years old can be dated quite well. In some cases, this has been of great help to the science of archeology which interests itself in man's ancient history. It has just about settled, for instance, the dates when Indians first arrived in various parts of North and South America.

This technique was developed by the American chemist, Willard Frank Libby, in 1947, and he received a Nobel Prize in 1960 in consequence.

Still another use for isotopes involves its killing properties, the very dangers I mentioned in the previous chapter. After all, gamma rays and high-energy particles can be deadly not only to human beings but to other forms of life as well. For instance, radioactive radiations will kill bacteria.

Now if food is exposed to radioactive radiations, the bacteria and other minute forms of life within it can be entirely wiped out. The food is made *sterile*. If it is kept sterile, the food will not spoil or discolor or lose its flavor. To be sure, we have other ways of sterilizing food. The usual method is to boil it for a period of time, then can it. The trouble with that is that boiling kills the bacteria but often changes the flavor.

Preservation of food by radioactivity may be the answer to the problem of keeping food indefinitely and retaining all the flavor of the fresh item.

So you see I have now listed a variety of different ways in which atoms can work for peace. I have described how isotopes can be helpful in:

1) Chemical and biochemical research
2) Medicine
3) Industry
4) Agriculture
5) Archeology
6) Home economics

That isn't the end, either. As isotopes become still cheaper and more plentiful, and as mankind learns more and more how to use them, there is no reason to suppose that isotopes won't become as widely useful as electricity. In fact, I haven't even yet mentioned the most important peacetime use of the atom — the production of atomic power.

Power A-Plenty

Energy can be wasted. It can be allowed to pass off simply as heat, accomplishing nothing (except perhaps to warm up a house). On the other hand, energy can be used to move objects against resistance. It can be used to lift an elevator against the force of gravity. It can cause a pile-driver to force long poles deeply into the ground against the resistance of the soil. It can cause negative and positive ions, against their own mutual attraction, to move into separate compartments in a storage battery.

Whenever energy causes any of these or similar things to happen, that energy is being converted into *work*. The rate at which the work is being done is called *power*.

Industry gets most of its power by using the energy that is released when coal is burned. Ships and locomotives used to get power by burning coal, but in recent years they have switched more and more to burning fuel oil. Automobiles and airplanes, as we all know, are powered by burning gasoline. The energy of falling water can also be used to supply power.

Even the power of atomic bomb explosions could be put to constructive use, instead of to simple destruction. Harbors might be blasted out, rivers turned out of their course, mountains bored through. The ground deep under the surface might be broken up to liberate new sources of oil and other minerals. The explosions might set up waves that could be used to study the interior.

Although none of this has actually come to pass yet, the energy produced by nuclear reactors has been turned into useful work of a quieter type. In other words, mankind is making constructive use of *atomic power*.

The most glamorous example is the *Nautilus,* an atomic-powered submarine launched in 1955. The *Nautilus* carries a nuclear reactor. This produces energy in the form of heat, which can be used to boil water, The steam so formed can by its pressure spin a turbine, after which the steam condenses back to water and is ready to be boiled again. The spinning turbine meanwhile is generating electricity. The electricity can then be used to run everything on the submarine from the engines to the jukebox.

A second atomic-powered submarine, the *Seawolf,* was added to the American Navy in 1956, and by the middle 1960's, the number of such submarines was nearing the hundred mark.

The special abilities of an atom-powered submarine, with its capacity for staying underwater indefinitely, was dramatically demonstrated in August, 1958, when the *Nautilus* crossed the Arctic Ocean, directly over the North Pole, remaining beneath the surface ice all the way.

The *Seawolf* remained underwater for 60 straight days in September and October of that year. In March, 1959, she surfaced at the North Pole.

Perhaps most amazing was the feat of the *Triton,* which is the longest submarine in the world at this moment and the only one to be powered by two nuclear reactors. In 1960, she circumnavigated the world underwater, following the route taken by Ferdinand Magellan four hundred and fifty years ago. Magellan's ships took three years to make the journey. The *Triton* took three months.

The Soviet Union has also built atomic-powered submarines and both it and the United States have built atomic-powered surface vessels.

Just the same, there are some difficulties involved in atomic power. In the first place, nuclear reactors must be quite large; otherwise the uranium would be below the critical size. In addition, the moderator takes up room. Besides the uranium and the moderator, there is *shielding*, perhaps in the form of concrete. (Concrete contains light atoms and isn't nearly as efficient as lead, for instance, so that it has to be made very thick. However, concrete is so much cheaper than metals that it is frequently used around nuclear reactors.) The shielding protects human beings from the radiation and the sub-atomic particles that are always being produced by a reactor. This adds to the size and mass of the reactors. Some of them are as big as five-story buildings.

As you see, then, while submarines and surface vessels and, possibly, large planes might be fitted out with nuclear reactors, smaller vehicles couldn't be. As far as we know today, there is no way of building a practical atomic-powered automobile. A nuclear reactor, complete with shielding, simply could not be squeezed under a car's hood. On the other hand, atom-powered spaceships are a distinct possibility some day.

Then there is the question of the uranium breakdown products formed during fission. Some of the breakdown products absorb neutrons strongly. If they were allowed to accumulate, they would take so many neutrons out of action as to stop the fission reaction entirely. (It would be like ashes finally choking out a fire.) For that reason, modern reactors aren't built in the pile form of the first University of Chicago reactor. Instead, the uranium is in the form of cylinders, or "slugs," which are inserted into holes in the moderator. Then, when breakdown products have accumulated, the slugs can be removed, the uranium purified, and new slugs formed and inserted, all by remote control.

In addition, the breakdown products are highly radio-active and must be buried somewhere to prevent harm to man. If a day comes when the earth is dotted with atomic-power plants, then probably the most important problem will be how to get rid of radioactive "ash," also called *atomic wastes,* safely.

Tests are being conducted to see if such atomic wastes might not be buried in unused salt mines or in deep rock layers. It is also possible that they may be melted into thick glass or ceramics and then dumped in the deep parts of the ocean.

On the more optimistic side, there is the fact that the atomic wastes still contain considerable energy. Such wastes have been used to power satellites, and may some day convert this energy to electricity. Such *atomic batteries* have been used to power satellites, and may some day power automobiles. Transuranium isotopes, such as curium-242, as well as the dangerous strontium-90, may be used as the power source. Thus, even these artificial elements, un-known in nature, may end as servants of man.

More Fuel

If atomic power is to become important, one thing that will be needed is uranium. Uranium is not a rare metal. There is more of it in the earth's crust than there is of copper, for instance, and you certainly have seen plenty of copper in your life. Uranium, however, is spread out pretty thin. There's never very much of it in any one place.

But uranium is now eagerly sought after, and "uranium strikes" are being made in many parts of the world.

It seems a shame, perhaps, that only seven atoms of uranium in a thousand are fissionable. Only seven atoms in a thousand are uranium-235, you remember. Well, there are ways of getting around that. Uranium-238 may not split when a slow neutron strikes it, but it can absorb that neutron and become uranium-239. Uranium-239, by losing two beta particles, becomes first neptunium-239, then plutonium-239. And plutonium-239 is fissionable!

It is possible, then, to build a nuclear reactor in such a way that while some neutrons keep the fission going, other neutrons form plutonium-239 out of uranium-238. This reactor produces more fissionable material, more *nuclear fuel,* than it uses up. It "breeds" fuel. Such reactors are called *regenerative reactors* or *breeders.* Regenerative reactors may also be designed to make use of thorium as fuel. This increases mankind's supply of nuclear fuel still further. The method of using thorium is to bombard thorium-232, the naturally occurring isotope, with neutrons, Uranium-233 is formed and it is this substance which is fissionable, as is its brother isotope, uranium-235.

The large quantities of plutonium made in this way may be used in A-bombs along with or instead of uranium-235. A nuclear reactor using only plutonium has been built. It doesn't require a moderator, but operates on fast neutrons. It is therefore called a *fast reactor.*

The Future

It must not be supposed that atomic power will be running the whole world next month, next year, or even next decade.

To be sure, the world's supply of petroleum, if we keep on burning it as fast as we do today, may be used up by the year 2000. That will be one major source of energy gone. However, the coal supplies of the earth ought still to last us several thousand years. And even though an ounce of uranium-235 will supply as much energy as 200,000 pounds of coal, coal still has certain advantages.

Coal is much more common than uranium, and is easier to handle. You can build a small coal fire, but you can't build a small nuclear reactor (at least not so far). Most important of all, coal is not radioactive, and neither is its ash. The dangers of nuclear reactors as a source of power are shown by the fact that there have been accidents in Canada and Great Britain which spread radioactivity about the neighboring areas. These accidents did not turn out to be serious, but they might have been. It shows the need for care and still more care. (Of course, people have been killed in coal mines and in oil explosions, too. Man has always lived with danger.)

For that reason atomic power did not replace power from other sources at once. It was too complicated, had a great many dangers and risks that had to be taken care of. Just the same people did begin to build atomic power plants to serve as a new source of energy.

First in the field as far as nuclear reactors for the production of electricity for civilian use was the Soviet Union. In June of 1954, they put into action a small power station with a capacity of 50,000 kilowatts.

By October, 1956, Great Britain had Calder Hall in operation. This is a nuclear reactor with a capacity of more than 90,000 kilowatts.

The United States was third. On May 26, 1958, Westinghouse completed a nuclear reactor for the production of civilian power at Shippingport, Pennsylvania, with a capacity of 60,000 kilowatts. In 1960, two more reactors were put into operation, one in Illinois and one in Massachusetts, with capacities of 180,000 kilowatts and 110,000 kilowatts, respectively. Others followed. Now the

United States produces many millions of kilowatts of atom-powered electricity.

The world is moving toward breeder reactors too. The Soviet Union puts its first large-scale breeder reactor into action in 1973. Great Britain and France are working on breeder reactors, too, and the United States plans its first by 1980.

Breeder reactors, which can use all the uranium and thorium in the world, can supply us with energy for a hundred thousand years or more. Even so the dangers of radioactive ash and of possible radiation leakage and disastrous explosions, remains so fission plants are not the entire answer.

There is much more energy to be obtained from hydrogen fusion. The United States, the Soviet Union, and Great Britain are working eagerly on methods that may serve to tame the H-bomb and turn its power to peaceful purposes.

Fusion power plants would have several advantages over fission power plants. The fuel would be deuterium. One pound of deuterium can produce 40 million kilowatt-hours and there is enough of it in Earth's oceans to last for billions of years. Of course, there is only 1 atom of deuterium for every 5,000 of ordinary hydrogen in water. Even so, 1 gallon of water contains enough deuterium to yield the power of 300 gallons of gasoline.

There would be no radioactive ash to get rid of since the products of fusion would not be radioactive. There is even reason to think that a fusion power plant, unlike a fission power plant, could not explode.

For fusion power plants to be practical, however, ways must be discovered to produce controlled temperatures of hundreds of millions of degrees without melting the power plant. At such high temperatures, the atoms of matter break down to a mixture of atomic nuclei and electrons. Such a mixture is called *plasma*. Stars such as our sun are composed of plasma.

Since plasma is composed of charged particles, it may be guided and controlled by the application of strong magnetic fields. Scientists are even now trying to find ways of holding the plasma within a field in such a way that it never touches any of the material containers inside which it flows. The plasma is pinched together by the magnetic forces and kept away from the container walls. This is called the *pinch effect*.

One of the devices in which scientists are trying to confine the plasma is called the *Perhapsatron* — the name coming from the notion that it might perhaps work. Another device is called the *Stellarator* because it is hoped that stellar temperatures may be reached there; that is, temperatures equal to those in the center of stars.

Another approach is to have a magnetic field stronger at the ends of the tube so that plasma is pushed back and kept from leaking. This is called a *magnetic mirror*.

The best device of this kind was worked out by Soviet scientists, and it was quickly adopted by Americans. It uses two magnetic fields and is called a *Tokamak* (which is an abbreviation of a long descriptive phrase in Russian).

Then, too, in 1960, a device called the *laser* was invented. This can put out a beam of light that can be concentrated on a tiny pinpoint. A great deal of energy can be delivered by laser, and if a number of large lasers all combine their energy on a tiny bit of frozen deuterium (mixed with frozen tritium so that fusion will start at a lower temperature) the problem may be solved.

There is considerable hope among scientists that nuclear fusion will be a practical source for electricity by 2000.

Meanwhile, practical applications of fusion research are

to be found. Plasma torches, emitting jets at temperatures up to 30,000 degrees C. in absolute silence, can far outdo ordinary chemical torches.

Nothing in the history of mankind has opened our eyes to the possibilities of science as has the development of atomic power. In the last two hundred years, people have seen the coming of the steam engine, the steam boat, the railroad locomotive, the automobile, the airplane, radio, motion pictures, television, the machine age in general. Yet none of it seemed quite so fantastic, quite so unbelievable, as what man has done since 1939 with the atom.

Man is confident, as never before, that, if he can only master himself, he can master the universe.

It was not many years ago that science-fiction writers (myself included) were considered queer because they wrote of atomic power and space travel. Now atomic power is a reality and rocket engines have sent satellites around the Earth, and out to Venus, Mars, Jupiter, and Mercury. Man himself has gone out into space and has left his footprints on the soil of the Moon.

What seems science fiction now will be hard fact tomorrow. Even after coal and uranium both run out, there is still the possibility of obtaining power from hydrogen fusion or directly from the sun's energy.

The sun, as we said earlier in the book, is a giant hydrogen bomb, pouring out unbelievable quantities of energy. That energy will last billions of years, and even the tiny fraction of it that strikes the earth could supply us with all the power we could use practically forever.

If only mankind can avoid destroying itself in atomic warfare, there seem to be almost no limits to what may lie ahead: inexhaustible energy, new worlds, ever-widening knowledge of the physical universe.

If only we can learn to use wisely the knowledge we already have . . .

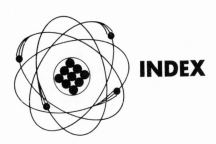

INDEX

207895

COALINGA STACKS
207895
539.7 ASI
Inside the atom.

539.7
A   Asimov, Isaac

              Inside the atom

DATE DUE

WEST HILLS COLLEGE LIBRARY
COALINGA, CALIFORNIA
DISCARD